Intentional Stance Towards Humanoid Robots

Agnieszka Wykowska
Editor

Intentional Stance Towards Humanoid Robots

Lessons Learned from Studies in Human-Robot Interaction

Springer

Editor
Agnieszka Wykowska
Istituto Italiano di Tecnologia
Social Cognition in Human-Robot Interaction Unit
Genova, Italy

ISBN 978-3-031-65482-4 ISBN 978-3-031-65483-1 (eBook)
https://doi.org/10.1007/978-3-031-65483-1

This Springer imprint is published by the registered company Springer Nature Switzerland AG
The registered company address is: Gewerbestrasse 11, 6330 Cham, Switzerland

Preface

With the re-birth of great interest in Artificial Intelligence and AI-empowered technologies, such as robots, we are all asking ourselves the question: what role will these new technologies play in our lives and in our societies? On the one hand, we seem to be excited about the new opportunities that AI and robots bring about. On the other hand, there are understandable fears: what if the AI becomes too powerful, what if the robots become too human-like?

In the context of the AI/robot-centred discussion, which is currently vibrant not only in scientific literature, but also in daily news and media, this book addresses the timely question: how do we, humans, approach robots? Do we see them as intentional agents that can potentially become our social companions? Or do we rather approach them as tools, man-made artefacts? What does it take for a robot to make an impression that it is an intentional/social agent? Is it its appearance? Its behaviour? Or, perhaps (and quite likely), the way it interacts with the human?

These questions are discussed here in the context of a theoretical framework of the concept of the Intentional Stance. Intentional Stance is a strategy we adopt when explaining and predicting the behaviour of other humans with reference to mental states: "she reached for this bottle of water because she *wanted* to drink, and she *believed* water would ease her thirst". "Wanting" and "believing" are mental states that we attribute to other humans, with the assumption that those are the underlying causes of the observable behaviour. Such predictive/explanatory strategy works well for humans, as humans are true intentional agents. Due to its success in predictive and explanatory power in interactions with other humans, we might have the tendency to use this strategy also in relation to other systems, such as humanoid robots. If that is the case, then adopting the Intentional Stance towards robots might have certain implications for how we approach robots in social contexts.

This book explores various possible factors that contribute to the tendency of adopting the Intentional Stance also to humanoid robots. Once the Intentional Stance is adopted, a robot might be treated more as an intentional (and possibly) social agent, rather than just a tool. It is thus crucial to understand what design principles can be provided to roboticists who aim to design robots that are perceived as intentional entities, and, in consequence, social companions. Although this is an exciting perspective, and possibly very beneficial for various application scenarios, such as childcare or elderly care, there are also some important ethical considerations to be made, and risks to be considered. Do we actually want to create robots that elicit the adoption of the Intentional Stance and that are treated as intentional agents? What if the line between the human/robot categories becomes blurred? These questions are addressed in the book, in the context of implications and consequences of adopting the Intentional Stance towards robots. One of these implications is potential moral responsibility. As intentional agents, should robots become also moral agents and legal *persons* in our societal and legal systems?

This book aims at offering the reader some answers to the above questions. The highlight is on *some* answers, because, as usually is the case, any empirical study yields more new open questions than answers. We hope that with some of the answers provided here, and new questions posed, the book will leave the reader with new reflections, and inspiration for scientific inquiry into the topic of attribution of intentionality to AI and robot technologies. We also hope that the inspiring conversations with prominent academics in the fields of Robotics (Hiroshi Ishiguro), Robot-Ethics (David Gunkel), Robotic Art (Bill Vorn), Human-Robot Interaction (Antonio Sgorbissa), Cognitive Science (Bertram Malle), and Philosophy (Daniel Dennett), which are included in this book, will provide the reader with comprehensive overview of scientifically and philosophically informed views on the following question: do we want to envisage a future in which robots are our daily companions?

Genova, Italy Agnieszka Wykowska

Acknowledgments

This work has received support from the European Research Council under the European Union's Horizon 2020 research and innovation programme, ERC Starting grant ERC-2016-StG-715058, awarded to Agnieszka Wykowska. The content of this book is the sole responsibility of the authors. The European Commission or its services cannot be held responsible for any use that may be made of the information it contains.

Contents

Contents

About the Editor

Agnieszka Wykowska is the head of the unit "Social Cognition in Human-Robot Interaction" at the Italian Institute of Technology (IIT), in Genoa, Italy. She is also the Coordinator of the Centre for Human Technologies, at the IIT. Her background is Neuro-Cognitive Psychology (Ludwig Maximilian University Munich, M.Sc. obtained in 2006) and Philosophy (Jagiellonian University in Kraków, M.A. obtained in 2001). She obtained a PhD in Psychology (2008) from the Ludwig Maximilian University Munich. She is Editor-in-Chief of the *International Journal of Social Robotics*. Between 2022 and 2024 she served in the role of President of the European Society for Cognitive and Affective Neuroscience (ESCAN). She is also board member of the Association of ERC Grantees and a delegate to the European Research Area (ERA) Forum—an EU expert group shaping EU science policies. Her research foci are interdisciplinary, bridging psychology, cognitive neuroscience, robotics and healthcare. She combines cognitive neuroscience methods with Human-Robot Interaction to understand the human brain mechanisms in interaction with other humans and with technology. In 2016 she was awarded the ERC Starting grant "InStance: Intentional Stance for Social Attunement", a project which spanned across six years and examined factors and conditions under which humans adopt the Intentional Stance towards humanoid robots. This book is a culmination of that project.

Contributors

Abdulaziz Abubshait Italian Institute of Technology, Social Cognition in Human-Robot Interaction Unit, Genova, Italy

Marwen Belkaid Italian Institute of Technology, Social Cognition in Human-Robot Interaction Unit, Genova, Italy

ETIS UMR 8051, CY Cergy Paris Université, ENSEA, CNRS, Cergy, France

Francesco Bossi Department of Information Engineering, University of Pisa, Pisa, Italy

Davide De Tommaso Social Cognition in Human-Robot Interaction Unit, Istituto Italiano di Tecnologia, Genova, Italy

Davide Ghiglino Italian Institute of Technology, Social Cognition in Human-Robot Interaction Unit, Genova, Italy

Kyveli Kompatsiari Italian Institute of Technology, Social Cognition in Human-Robot Interaction Unit, Genova, Italy

Department of Applied Mathematics and Computer Science, Cognitive Systems, Technical University of Denmark, Kongens Lyngby, Denmark

Serena Marchesi Italian Institute of Technology, Social Cognition in Human-Robot Interaction Unit, Genova, Italy

Uma Prashant Navare Italian Institute of Technology, Social Cognition in Human-Robot Interaction Unit, Genova, Italy

ETIS UMR 8051, CY Cergy Paris Université, ENSEA, CNRS, Cergy, France

Ziggy O'Reilly Italian Institute of Technology, Social Cognition in Human-Robot Interaction Unit, Genova, Italy

Department of Psychology, Università di Torino, Turin, Italy

Lorenzo Parenti Italian Institute of Technology, Social Cognition in Human-Robot Interaction Unit, Genova, Italy

Department of Psychology, Università di Torino, Turin, Italy

Jairo Perez-Osorio Italian Institute of Technology, Social Cognition in Human-Robot Interaction Unit, Genova, Italy

Cognitive Psychology & Ergonomics, Technische Universität Berlin, Berlin, Germany

Cecilia Roselli Italian Institute of Technology, Social Cognition in Human-Robot Interaction Unit, Genova, Italy

Nicolas Spatola Artimon Perspectives, Paris, France

Agnieszka Wykowska Istituto Italiano di Tecnologia, Social Cognition in Human-Robot Interaction Unit, Genova, Italy

A special bonus contribution from the following experts who shared their insights in interviews included in this book

Daniel Dennett Center for Cognitive Studies, Tufts University, Medford, MA, USA

David Gunkel Department of Communication, Northern Illinois University, DeKalb, IL, USA

Hiroshi Ishiguro Department of Systems Innovation, Osaka University, Osaka, Japan

Japan & Advanced Telecommunications Research Institute, Kyoto, Japan

Bertram Malle Department of Cognitive, Linguistic, and Psychological Sciences, Brown University, Providence, RI, USA

Antonio Sgorbissa Department of Computer Science, Bioengineering, Robotics, and Systems Engineering, University of Genova, Genova, Italy

Bill Vorn Department of Studio Arts, Concordia University, Montreal, Canada

Chapter 1
Introduction

Agnieszka Wykowska

1.1 Introduction

Let me start by a quotation taken from the *Hitchhiker's Guide to the Galaxy* by Douglas Adams. It is a passage in which Marvin, an ***android robot***, carries out a short exchange with Ford Prefect:

> *'That ship hated me', he* [Marvin] *said dejectedly, indicating police craft.*
> *'That ship? said Ford in sudden excitement'. 'What happened to it? Do you know?'*
> *'It hated me because I talked to it.'*
> *'You talked to it?' exclaimed Ford. 'What do you mean, you talked to it?'*
> *'Simple. I got very bored and depressed, so I went and plugged myself in to its external computer feed. I talked to the computer at great length and explained my view of the Universe to it', said Marvin.*
> *'And what happened?' pressed Ford.*
> *'It committed suicide', said Marvin.*[1]

No matter how science fiction this passage seems to be, there is an aspect to it that might not be as science fiction as it appears. Although ships committing suicide (due to being overwhelmed by reflection over the universe) and depressed robots are indeed rather in the realm of fiction, the *attribution* of feelings or thoughts to the

[1] Adams, D. "Hitchhiker's Guide to the Galaxy" (1979 edition). Pan Books, Pan Macmillan Ltd, London, p. 179.

A. Wykowska (✉)
Istituto Italiano di Tecnologia, Social Cognition in Human-Robot Interaction Unit, Genova, Italy
e-mail: agnieszka.wykowska@iit.it

A. Wykowska (ed.), *Intentional Stance Towards Humanoid Robots*,
https://doi.org/10.1007/978-3-031-65483-1_1

machines might actually be real. At least in some contexts and under some circumstances.

People have a strong tendency to anthropomorphize all sorts of non-living things—from plush toys, cartoon characters to cars, machines, computers, and robots. The process of anthropomorphism is ubiquitous and so omnipresent that it almost appears as the most natural thing that humans do with respect to various non-living things. Anthropomorphism is defined as *'the interpretation of nonhuman things or events in terms of human characteristics'* (Guthrie, 2008). Interestingly, the same source later describes: *'Most scholars since the time of the English philosopher Francis Bacon (1561–1626) have agreed that the tendency to anthropomorphize hinders the understanding of the world, but it is deep-seated and persistent.'* Anthropomorphism does indeed hinder understanding of the world, as it entails attribution of certain illusory and unreal qualities to things and events. Think of seeing human faces in clouds or car fronts, or explaining a thunderstorm with reference to an ancient god 'being angry' with humans.

1.2 The Intentional Stance

What is it, then, that makes us 'see' humans in entities that are not human? An important aspect of anthropomorphism is not necessarily related to how the entity looks and appears in terms of physical characteristics (although these are crucial features), but rather, to what the entity is doing and what are its motivations for certain behaviours and actions. In fact, we might anthropomorphize things not so much because we see human-like appearance in them, but rather because we would like to be able to understand their behaviour in a human-like manner. This might make it easier for us to be around them, and indeed, to explain and predict how they are going to behave. But how do we explain and predict behaviours in a human-like manner? In other words, how do we explain the behaviour of other humans?

According to Daniel Dennett (1987), when we observe other humans, we adopt the so-called ***Intentional Stance***. We refer to others' mental states, such as *beliefs, desires,* and *intentions,* to understand, explain, and predict their behaviours. In more general terms, Dennett proposed that different strategies (or stances) work best for different systems (Dennett, 2003). For example, when driving a car, the driver predicts that the car will reduce speed when the brake pedal is pushed. In this example, a *Design* Stance is the most successful, as the best (or most efficient) predictions are made when one refers to how the system has been designed to behave. In the case of other human agents, the *Intentional Stance* works best. For example, when I see my best friend extending her arm with a glass of water in my direction, I assume that she *intends* to hand me that glass of water because she *believes* that I am thirsty and she *wants to* ease my thirst. By the same token, when I see somebody pointing to an object, I try to infer why they do so and what it is that they want to show me. These examples clearly illustrate that Intentional Stance is the most natural way for us to explain other humans' actions and behaviours.

However, going back to anthropomorphizing non-human entities, it is also likely that we sometimes, if not often, adopt the Intentional Stance towards artefacts. Especially if they look like humans, and perhaps behave in a seemingly human-like manner. Think of humanoid robots. Why is that the case?

1.3 The Intentional Stance Towards Humanoid Robots

Humanoid robots are a category of artefacts that have been built to resemble humans in their physical characteristics (they typically have human-like artificial 'body', namely a torso, a head, two arms, and two legs, though some humanoid robots may replicate only part of the body, for example, only from the waist up). Such design might have been chosen for functional reasons, as human-like shape allows for better navigation and functionality in human-centred environments (think of, for example, using legs, as opposed to wheels, to climb up stairs; or navigating between densely arranged tables in a restaurant while manipulating objects). Interestingly for the purposes discussed here, human-like shape might also be a deliberate design choice with the aim of making humanoid robots good candidates for being social companions of humans. If a robot looks like a human and walks like a human, it might be more likely that it is perceived as a social entity. Having eyes like a human or lips that can smile, or express sadness, makes a humanoid robot much more likely to evoke social attitudes towards it, as opposed to a robotic manipulandum without any social 'expressiveness'. Given that humanoid robots do resemble humans in the physical appearance and the way they act upon the environment, it might be quite tempting to not only see them as social entities, but also to adopt the Intentional Stance towards them. That is because our brain tends to automatically anthropomorphize things that look or behave like a human.

Importantly, as suggested in the passage above, the human-like shape endows a robot also with human-like movement repertoire. And, as mentioned earlier, it is often the human-like behaviour that is the key to adopting the Intentional Stance. The seminal study by Heider and Simmel (1944) showed that even simple geometrical figures (squares and triangles), when displaying certain patterns of movements on the screen, are capable of evoking the adoption of Intentional Stance in the observers.

It seems that our brain's mechanisms, which have developed for interactions with our conspecifics, might be 'hijacked' by other types of entities that resemble humans, especially if they also behave and act like humans (Wykowska, 2020; Wykowska et al., 2016). If a robot walks like us, grasps objects like us, and moves its eyes the way we do, we might be inclined to think that it actually does so, out of the same motives and reasons that we do. That is, as a consequence of mental states.

Having said this, the answer to whether we do, or do not, adopt the Intentional Stance towards humanoid robots is not straightforward. Certainly, the human-like behaviour (and appearance) of a humanoid robot plays a crucial role. But there are several other factors that are also important to consider when addressing the question of whether and when humans adopt the Intentional Stance towards robots.

1.4 Why Bother? Why Do We Want to Understand If, and When, Humans Adopt the Intentional Stance Towards Humanoid Robots?

Understanding the factors contributing to the likelihood of adoption of the Intentional Stance towards non-human artificial agents is of great scientific interest, as it informs about the external features (appearance and behaviour) of an agent that manifests intentionality. Thus, this addresses the relationship between the internal property of being an intentional *thinking* agent and the way this can be inferred from external characteristics. This problem is analogous to the issue of the Turing test (Turing, 1950), where the idea is to infer intelligence from external manifestations thereof, evident through interaction.

However, the topic of Intentional Stance towards humanoid robots is not only of scientific interest but has also practical implications. Humanoid robots are currently in the limelight not only within the scientific community, but also among the general public. Humanoid robots offer a promise of being our future companions, assistants in daily mundane tasks, and of being at service for not only the dirty dangerous jobs such as search and rescue in disaster recovery, but also in daily factory settings and home environments.

However, to make robots work efficiently hand-in-hand with humans, it is of utmost importance that they are well 'attuned' to how humans work and operate. Think of any type of coordination task between two humans, such as carrying a piece of furniture together or cooking a meal together. To perform these kinds of tasks well with others, we need to be able to read out their intentions ('is she taking the knife to cut the tomato or rather to slice the onion?'), coordinate movements together ('she is tilting the table that we are carrying together to the left, so I need to tilt in the same direction'), and to engage in joint attention, meaning to attend where our interaction partner is attending ('she is looking at the tomato, so I need to attend there, too, to monitor what she is going to do with it next').

All these mechanisms can be grouped under one umbrella of ***social attunement***. How does Intentional Stance, however, relate to social attunement? Think of the example given above of an interaction partner extending her arm with a glass of water and the observer assuming as that the person *intends* to hand over that glass of water. This example clearly indicates that adopting the Intentional Stance is crucial for social attunement, through affecting fundamental processes of social cognition, such as joint attention. We will engage in joint attention only if we believe that there is an actual intention behind the interactor's action. Therefore, we need to first assume intentionality to engage in a joint attention. It makes sense to attend to where our interaction partner directs our attention (through pointing or gazing) because such behaviour carries socially relevant (and often communicative) content. In contrast, when a mindless machine is pointing to a given direction, it might make less sense to attend there, as the behaviour does not carry any meaning. As joint attention is a basic component of social attunement, as well as of any type of joint action and coordination, it might be the case that without adoption of the Intentional Stance, coordination or any sort of joint task will not be performed efficiently.

Another reason why adopting the Intentional Stance towards robots might be important is ***social companionship***. Think, for example, of an elderly person whose robot companion reminds them to take specific medication at a specific time of the day. If the robot is endowed with trust and intentionality, the person might be more inclined to follow its advice. If an elderly person adopts the Intentional Stance towards their robot companion, the robot's advice is taken as a result of a good intention or willingness to help, rather than just as pre-programmed 'alarm clock', which is much more tempting to ignore.

Finally, last, but not least, eliciting the adoption of the Intentional Stance, and, in consequence, better social attunement might be crucial in the context of ***educational robotics***. Robots that are to assist in educational activities might certainly be more efficient when children see them as intentional agents (and social companions), as opposed to only mundane artefacts. Attribution of intentionality might increase long-term engagement in educational activities and might prevent abandoning the 'mechanical toy' after a few instances of interactions. Assuming, of course, that the robot will be endowed with sophisticated enough artificial intelligence to maintain the educational activities interesting and engaging enough over the course of a longer-term interaction.

1.5 Ethical Considerations

Having discussed the positive reasons why adopting the Intentional Stance towards humanoid robots might be beneficial for humans, one needs to think also of the potential negative consequences of eliciting the adoption of Intentional Stance.

First, deception: robots do not have mental states. Therefore, eliciting in human observers the illusion that they operate on the basis of feelings, desires, and intentions might not be ethical. This is a serious problem that needs to be considered in the context of research on Intentional Stance towards robots.

Second, attachment: vulnerable individuals (children, the elderly) might become over-attached to robots which create the impression of being intentional. This might result in social withdrawal with respect to true social interactions with other humans.

Third, obscuring the boundaries between what is human and what is not: It is of utmost importance to keep the boundaries between who is a natural human (intentional agent) and what is just an artefact. Eliciting the adoption of Intentional Stance towards a robot might blur this boundary. Think again of Marvin the depressed robot from '*Hitchhiker's Guide to the Galaxy*.

While all these critical ethical aspects need to be taken very seriously, it is important to note that the scientific community first needs to understand when, and under what conditions, people actually adopt the Intentional Stance, in order to later take ethical measures that prevent negative consequences. We first need to understand what type of design of a robot might potentially elicit the Intentional Stance before we can control, or legally regulate, this kind of design for ethical reasons (assuming that it will be deemed necessary).

In this context, this book provides some initial insights into the factors that play a role in adopting the Intentional Stance towards robots, leaving most of the ethical discussion for later, for the stage when humanoid robots actually become produced at a commercial scale (if ever) and when ethical decisions need to be made regarding which design is ethically correct and for what sort of application. As of today, the robots produced in research laboratories are still too immature in terms of technology and artificial intelligence, with which they are endowed, to be perceived as intentional agents by anyone in the real world.

1.6 Content of the Book: Factors Affecting the Adoption of Intentional Stance Towards Humanoid Robots

This book is primarily based on research results of a 6-year project titled 'Intentional Stance for Social Attunement' (InStance), funded by the European Research Council and awarded to Agnieszka Wykowska, carried out at the Italian Institute of Technology from 2017 until the end of 2022. Throughout these 6 years, the team of Agnieszka Wykowska have addressed the different factors that influence the likelihood of adopting the Intentional Stance towards robots. The chapters of the book are organized along specific factors that were the focus of exploration throughout the project. Each chapter is authored by team members whose research centred around that given theme within the project. Before discussing those factors, however, one needs to first understand how to operationalise the philosophical concept of the Intentional Stance, in such a way that it would be empirically measurable whether a given person adopts (and to what extent) the Intentional Stance towards another agent.

- With this in mind, the next chapter (Chapter 2) addresses the issue of ***operationalization of the Intentional Stance***. To operationalise the concept of the Intentional Stance, however, one needs to first define the concept well and distinguish it from similar, but not identical, concepts, such as anthropomorphism or Theory of Mind. Chapter 2 first provides a delineation of the concept of the Intentional Stance and then discusses how the various similar concepts have been operationalized and empirically addressed in literature. The chapter concludes with an in-depth description of tools and measures (both behavioural and neural) of Intentional Stance. Chapter 2 is then followed by chapters which describe empirical investigations of various factors that influence (or potentially influence) the adoption of the Intentional Stance towards humanoid robots, namely:

- ***The Human***. Humans do not approach an interaction with a robot (or with any other entity) as blank slates. We have certain ideas about the interactant, what category of an entity it is, and how we expect it to behave. Perhaps we have seen a sci-fi movie depicting a robot. This would give us some preconceptions about what to expect from a robot. Or perhaps we have been culturally endowed with the idea that not only humans, but also other living beings and artefacts have a 'spirit' and, potentially, a 'mind' of their own. This would make us likely to adopt the Intentional Stance towards a robot. We might also have our individual tendencies to interpret the behaviour of other entities with reference to mental states. On the other hand, however, we might have received a certain type of education (e.g. in robotics) that clearly prevents us from attributing mental states to robotic artefacts, and, thus, reduce the likelihood of adopting the Intentional Stance towards a robot. In sum, humans have their individual, cultural and educational backgrounds, as well as the priors and preconceptions with which they enter an interaction with the robot. This will determine their likelihood of adopting the Intentional Stance towards the robot, as discussed in Chapter 3.
- ***The Robot***. As discussed above, the robot's appearance and behaviour are likely candidates to affect the likelihood of adopting the Intentional Stance. Think of the C-3PO robot depicted in *Star Wars*. The way it looks—even though made of metal—automatically elicits anthropomorphism due to its human-like shape. Its human-like conversations, gait, and movement repertoire only further reinforce the impression that its actions are governed by thoughts, feelings, and intentions. On the other hand, a robotic rover on Mars is quite unlikely to evoke in us the tendency to explain its behaviour in mentalistic terms. And that is partially due to its appearance, and partially because it moves around Mars on wheels and 'perceives' the world through its cameras that do not resemble human eyes. Taken together, the likelihood of adopting the Intentional Stance is to a large part determined by how a robot 'presents itself' to the human observer. This is discussed in Chapter 4. Note that the focus of the book is on nonverbal behaviours. This is because of the book's particular interest is how embodiment (and, in consequence, its motor repertoire) affects adoption of the Intentional Stance. As verbal (and language-mediated) communication can occur also with disembodied agents (think of Alexa, Siri, or ChatGPT), Chapter 4 does not analyse robots' verbal behaviour, as the interest is in how the body can convey and manifest intentionality.
- ***The Interaction***. One cannot neglect the importance of the interaction itself. The likelihood of adopting the Intentional Stance is also modulated by the interaction context. If the interaction promotes social bonding, creates an opportunity of 'being together' in the context of sharing a task, or evokes synchronization in reactions to the events occurring in the external world, the human interactant might have the impression that the robot is 'like me' as it shares 'my way' of seeing the world, acting within it, and responding to its events. Thus, if the robot is 'like me', the human is more likely to adopt the Intentional Stance towards it,

as s/he also adopts the Intentional Stance towards him/herself. Hence, the interaction might be a factor (over and above of what the interactants bring into the game) that influences if, and to what extent, we adopt the Intentional Stance towards our robotic interaction partners. This theme is discussed in detail in Chapter 5.

- ***Culture.*** Apart from the individual differences, such as individual traits, biases, and education, also cultural background is a key factor that might influence the likelihood of adopting the Intentional Stance. From aspects such as collectivism versus individualism, which might affect adopting the Intentional Stance towards robots due to a higher focus on social interactions (collectivism) versus emphasis on self-reliance (individualism); to historical and religious traditions of various cultures that might bias towards different degrees of attribution of mental states to artefacts (think of, for example, the Shinto religion which promotes animism; in contrast to the Christian tradition that is based on the mind–body dualism and high degree of anthropocentrism). Given that such cultural 'biases' are deeply engrained in our attitudes towards the world and towards others, it is likely that they might affect the adoption of the Intentional Stance towards robots. Chapter 6 is dedicated to the topic of culture as a factor influencing the adoption of the Intentional Stance. It discusses the issue both from the theoretical and the empirical side.
- ***Intentional Stance versus Moral Stance.*** After having discussed the factors that contribute to the adoption of the Intentional Stance, the following chapter (Chapter 7) discusses the *consequences* of adopting the Intentional Stance in terms of whether we perceive robots as potential *persons* with moral rights and moral responsibilities. This has obvious implications for legislation and how we attribute moral (and legal) responsibility to autonomous robots that might in the near future coexist with us in our societies and legal frameworks.
- ***Interviews.*** Each chapter of the book is followed by a **bonus material**: **an interview** with a selected prominent researcher or academic whose work is centred around the topic of the given chapter. Thus, the chapter which discusses the 'Human' factor in adopting the Intentional Stance is followed by an interview with ***Prof. Bertram Malle,*** whose research has been dedicated to examining social cognition and attribution of intentionality and moral psychology in the context of Human–Robot Interaction. The chapter discussing the 'Robot' (its appearance and behaviour) is followed by an interview with ***Prof. Hiroshi Ishiguro,*** who has devoted his professional life to building androids—robots that are extremely human-like in appearance. The chapter on 'Interaction' is followed by an interview with an artist, ***Bill Vorn***, who creates robotic art installations and performances, and who shared his insights on social attunement between his robot installations and the observers, from a unique, artistic perspective. The chapter on 'Culture' is followed by an interview with ***Prof. Antonio Sgorbissa***, whose recent work focuses on designing culturally competent social robots. In the final interview of the book, as a special highlight, ***Prof. Daniel Dennett*** himself, the originator of the Intentional Stance concept, shared his insights on the

idea of adopting the Intentional Stance towards robots—the promises and dangers in our future world, where robots will likely become our assistants and companions.

1.7 Introduction of the Main Protagonist of This Book: The iCub Humanoid

Before moving to the next chapter of the book, a brief introduction of the iCub humanoid robot (Metta et al., 2010) (Fig. 1.1) is needed. iCub accompanied the team of Agnieszka Wykowska thoroughout the Intentional Stance project. Most of the work presented in this book refers to iCub—the robot involved in the Human–Robot Interaction protocols. iCub is also *the* robot depicted in the Intentional Stance (InStance) Test, a tool developed to operationalize the highly philosophical concept

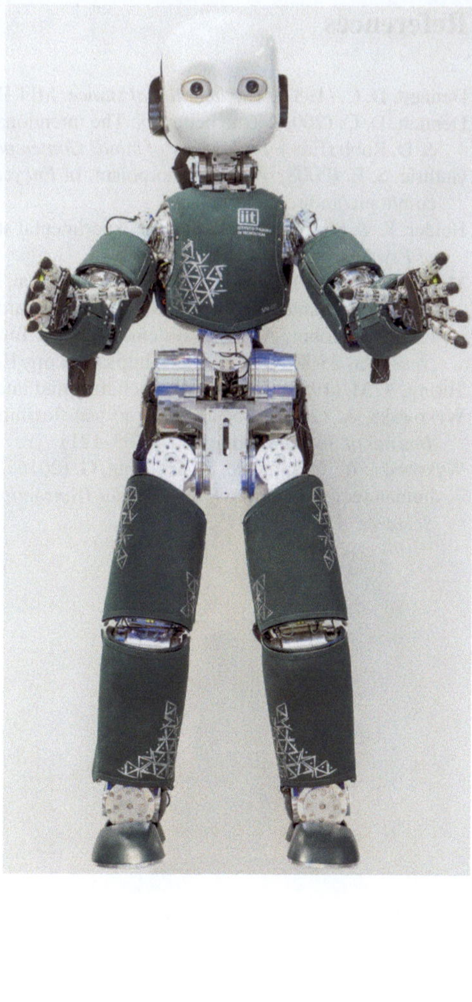

Fig 1.1 The iCub humanoid robot, designed and developed at the Italian Institute of Technology (IIT)

and to 'measure' the degree to which people adopt the Intentional Stance (towards iCub) (see Chapter 2).

iCub is a humanoid robot designed and developed at the Italian Institute of Technology. iCub has been conceived as a platform to conduct research on embodied artificial intelligence—with the idea to develop and test computational models of perception, cognition, and action. iCub is 104 cm tall, and this approximates the height of a 5-year-old child. It also has a child-like appearance through the design of its head and face where its large eyes are perhaps the most salient features on the otherwise rather plain white-plastic mask. iCub's hands have been developed to have a rich degree of dexterity and sophisticated manipulation capabilities, thereby affording Human–Robot Interaction tasks with joint manipulation. iCub can also walk and crawl, and is endowed with sensorized full-body skin which allows it to receive haptic information related to its body. iCub is an open-source platform, both in terms of software and in terms of hardware. For more information on the iCub platform, the reader is invited to visit https://icub.iit.it.

References

Dennett, D. C. (1987). *The intentional stance*. MIT Press.

Dennett, D. C. (2003). True believers: The intentional strategy and why it works. In T. O'Connor & D. Robb (Eds.), *Philosophy of mind: Contemporary readings* (pp. 370–390). Routledge.

Guthrie, S. E. (2008). Anthropomorphism. In *Encyclopaedia Britannica*, https://www.britannica.com/topic/anthropomorphism

Heider, F., & Simmel, M. (1944). An experimental study of apparent behavior. *American Journal of Psychology, 57*, 243–259.

Metta, G., Natale, L., Nori, F., Sandini, G., Vernon, D., Fadiga,L., von Hofsten, C., Rosander, K., Lopes, M., Santos-Victor, J., Alexandre, Bernardino, A., & Montesano, L. (2010). The iCub humanoid robot: An open-systems platform for research in cognitive development. *Neural Networks, 23*(8–9), 1125–1134. https://doi.org/10.1016/j.neunet.2010.08.010

Turing, A. M. (1950). Computing machinery and intelligence. *Mind, 59*, 433–460.

Wykowska, A. (2020). Social robots to test flexibility of human social cognition. *International Journal of Social Robotics, 12*, 1203–1211.

Wykowska, A., Chaminade, T., & Cheng, G. (2016). Embodied artificial agents for understanding human social cognition. *Philosophical Transactions of the Royal Society London: B. Biological Sciences, 371*, 20150375.

Chapter 2
Delineation and Operationalisation of the Concept of the Intentional Stance

Serena Marchesi, Nicolas Spatola, and Agnieszka Wykowska

From the perspective of others, I am part of the world, but when I observe the world from my perspective I am nowhere to be found. To observe is to create perspective. I can never cease to be the point of origin from my perspective. I observe that which is not myself. This is the first principle to find. Ergo Proxy.

– Shūkō M., Ergo Proxy (2006) Manglobe.

2.1 Introduction

Artificial agents, such as humanoid robots, can be viewed as technologically opaque entities (Surden & Williams, 2016). Their complexity, both in terms of hardware and software, places them in an ambiguous state, where they are unmistakably man-made creations, yet they possess the potential to be perceived as social actors. Some scholars even propose that this dual nature of humanoid robots might establish a new ontological category (NOC) (Kahn & Shen, 2017), making the exploration of people's attitudes towards them a subject of great interest for social and cognitive scientists. One of those attitudes is the Intentional Stance—a stance taken towards humanoid robots when explaining and predicting their behaviour.

S. Marchesi (✉) · A. Wykowska
Istituto Italiano di Tecnologia, Social Cognition in Human-Robot Interaction Unit, Genoa, Italy
e-mail: agnieszka.wykowska@iit.it

N. Spatola
Artimon Perspectives, Paris, France

A. Wykowska (ed.), *Intentional Stance Towards Humanoid Robots*,
https://doi.org/10.1007/978-3-031-65483-1_2

2.2 Conceptual Delineation

The aim of this chapter is to describe means of operationalisation of the concept of the Intentional Stance. However, to operationalize a concept, one first needs to make a clear conceptual differentiation between the target concept and related/similar concepts that can often be confused. In the context of the Intentional Stance, those are anthropomorphism and the Theory of Mind.

2.2.1 Anthropomorphism

Anthropomorphism is a broad concept that can be generally defined as the tendency to attribute human traits, emotions, or intentions to non-human entities (Epley et al., 2007; Waytz et al., 2010a, b). Anthropomorphism is the first factor that contributes to the perception of artificial agents as potential social partners. There is evidence from developmental psychology that the tendency to anthropomorphize non-human beings is present in humans from an early age. Interestingly, the tendency to anthropomorphize is associated with a high tendency to apply egocentric viewpoint—from seeing faces in car fronts or the moon, to applying egocentric reasoning (i.e., the belief that others have the same experiences, perceptions, and thoughts as we do) (Barrouillet, 2015). Since humans have the tendency to extrapolate these characteristics to other humans and non-humans, it is generally agreed that anthropomorphism is a pervasive cognitive bias (Airenti et al., 2019; Andrews & Huss, 2014; Dacey, 2017; Epley et al., 2007; Jones, 2022).

To better understand anthropomorphism we may emphasise its theoretical history. Since the early twentieth century, different theories have been proposed to describe the anthropomorphism process (Table 2.1).

Among these theories, the most prominent remains the three factors model from Epley et al. (2007), which argues that anthropomorphism is grounded in three factors:

Elicited agent knowledge refers to the knowledge or information that is activated or accessed by an individual when encountering non-human agents. It is the base of knowledge that individuals draw upon when reasoning about the properties, characteristics, and mental states of non-human entities. The concept of elicited agent knowledge suggests that when people encounter non-human agents, they rely on their existing knowledge and representations of humans, or self, to make inferences about these non-human entities. This knowledge can be acquired through direct or indirect experiences with humans, as well as through cultural and social influences. Elicited agent knowledge can encompass various aspects, including knowledge about human behaviour, emotions, intentions, and mental states. It is the cognitive framework that individuals use to understand and interpret the actions and attributes of non-human agents based on their understanding of humans or themselves. The accessibility and applicability of elicited agent knowledge play a crucial role in

Table 2.1 Conceptualisation of anthropomorphism across literature

Authors	General process of anthropomorphism
Piaget (1929)	Anthropomorphism would rely on an egocentric reasoning in childhood.
Heider and Simmel (1944)	When objects are moving without any identifiable cause, there is a tendency to interpret the movements as intentional (i.e. anthropomorphic).
Fisher (1991)	Two ways of anthropomorphism: Interpretative anthropomorphism as the attribution of intentions, beliefs, and emotions to non-human agents based on their behaviour; Imaginative anthropomorphism as the representation of imaginary and fictional characters as human-like.
Mithen and Boyer (1996)	Anthropomorphism results from the interaction between social intelligence, processing social information, and a mechanism processing biological information.
Caporael and Heyes (1997)	Anthropomorphism relies on a cognitive default system, which is restrained when alternative explanations appear more suitable to explain or describe non-human actions.
Caporael and Heyes (1997)	Anthropomorphism relies on inter-species behaviour recognition.
Guthrie (1997)	Anthropomorphism relies on a cognitive default system to interpret ambiguous stimulus in the environment as human-like.
Epley et al. (2007)	Schemas about humans are used as the basis for explaining other entities, because this knowledge is more accessible and more detailed than knowledge about non-human entities. This process is moderated by three factors: Elicited agent knowledge, that is, the amount of prior knowledge held about an object and the extent to which that knowledge is accessible; Effectance, that is, the willingness to interact and understand the environment; Sociality, that is, the willingness to establish social connections.
Urquiza-Haas and Kotrschal (2015)	Anthropomorphism relies on a non-reflective and a reflective process. The non-reflective process would be automatic and less affected by cultural differences, while the reflective process would be more prone to inter-individual differences.
Dacey (2017)	Intuitive anthropomorphism is a heuristic (cognitive bias) used by our unconscious (folk) psychology to understand non-human animals.
Airenti (2018)	Anthropomorphism is grounded in interaction. In interaction, a non-human entity assumes a place that generally is attributed to a human interlocutor. This approach is based on four main assumptions: Adults under certain circumstances may anthropomorphize entities even if they know that these entities have no mental life; Anthropomorphism is situational and does not depend on a specific target; There is no consistency among the entities that are anthropomorphized; Inter-individual variability in anthropomorphism is a result of affective states rather than of different degrees of knowledge about the target.
Spatola et al. (2022)	Anthropomorphism relies on a default social cognition system that could be bypassed by an active process when sufficient cognitive resources are available. This would result in a switch to a physical cognition system favouring target-specific information and, concomitantly, restricting anthropomorphic inferences (more accessible).

anthropomorphism. If individuals have readily accessible and applicable knowledge about humans or self, they are more likely to attribute human-like characteristics, traits, and mental states to non-human agents.

Sociality relates to the motivation and desire for social connection with others. It is the inherent human need to establish and maintain relationships and affiliations with other individuals. In the context of anthropomorphism, sociality plays a crucial role in shaping the tendency to attribute human-like characteristics, intentions, and emotions to non-human agents. The theory suggests that when individuals experience social disconnection or isolation, they are more likely to anthropomorphize non-human agents as a means of seeking social connection. Anthropomorphizing non-human agents allows individuals to create a sense of social bond and connection, even if it is with entities that are not human. This social bond can alleviate the pain of social disconnection and fulfil the need for social interaction. Furthermore, sociality motivation influences the perception and interpretation of non-human agents. When individuals are motivated to establish and maintain social connections, they are more likely to attribute human-like traits and behaviours to non-human agents.

Effectance motivation depends on the inherent human desire to interact effectively with the environment by increasing comprehension, predictability, and control. It is the motivation to understand and make sense of non-human agents by attributing human-like characteristics, intentions, and behaviours to them. Effectance motivation plays a crucial role in anthropomorphism, as it drives individuals to seek comprehension and predictability in their interactions with non-human agents. When faced with uncertainty or ambiguity regarding the behaviour or nature of a non-human agent, individuals with high effectance motivation are more likely to activate anthropomorphic representations as a means to reduce uncertainty and increase their sense of control. Effectance motivation influences anthropomorphism by enhancing individuals' ability to explain and predict the behaviour of non-human agents. By attributing human-like traits and intentions, individuals can make sense of the actions and behaviours of non-human agents, reducing the complexity and uncertainty of their interactions. Furthermore, effectance motivation can lead individuals to apply anthropomorphic representations to non-human agents in order to increase their own efficacy and competence in interacting with them. By perceiving non-human agents as more human-like, individuals may feel more capable of understanding and effectively engaging with these agents.

Overall, anthropomorphism is a very broad concept denoting the human tendency to attribute (or project) human traits and characteristics to non-human entities. Intentional Stance is a very closely related concept, but not identical, to the concept of anthropomorphism. We propose that anthropomorphism is a broader concept which encompasses, but is not limited to, the concept of the Intentional Stance. This is because anthropomorphism relates not only to the tendency of explaining and predicting behaviour with reference to mental states, but it also relates to attributing human-like features in physical appearance (seeing 'face' in the moon) or perceiving general human-likeness in an entity. In fact, Spatola and Marchesi investigated the relationship between anthropomorphism and the adoption of the Intentional Stance in two separate studies, revealing that while they rely on similar mechanisms, Intentional Stance and anthropomorphism are distinct concepts (Marchesi et al., 2021, 2022). Anthropomorphism involves attributing human-like qualities or characteristics to non-human entities, such as animals or robots, encompassing physical features, emotions, thoughts, intentions, behaviour, or expressions projected onto them. On the other hand, the Intentional Stance is more specific, involving the attribution of beliefs, desires, and intentions to other entities, including robots. Thus, the adoption of the Intentional Stance is a form of anthropomorphism, but anthropomorphism itself encompasses a broader range of projections of human traits onto non-human entities, to make them more relatable or familiar.

2.2.2 *Theory of Mind*

From the very early stages of development, we learn to reason about the motives and reasons underlying the observed behaviour of others. We learn to search for those motives among the mental states that we attribute to others: 'S/he returned home shortly after leaving, because she *forgot* her wallet and she *believes* she left it home'. The ability to reason about others' mental states—which are assumed to be the underlying causes of behaviour—is crucial for daily functioning in the social environment and is critical during development: it provides foundations for language acquisition, allows differentiating between self and others, and is indispensable in social interactions. Interestingly, representing the mental states of others occurs effortlessly, automatically, and unconsciously. This ability of referring to others' mental states in explaining their behaviour has been termed *mentalizing* (Frith & Frith, 2006) or using a Theory of Mind (ToM: Baron-Cohen, 1997; Premack & Woodruff, 1978).

The human ToM has been extensively studied over the past decades. Initial cognitive approaches by Heider (1958), for instance, suggested that people have a general understanding of others' ideas and actions in particular situations—a 'common-sense or folk psychology' that helps them deal effectively with social situations. Subsequently, Premack and Woodruff (1978) defined ToM as the ability to reason about other's behaviour and mental states. Baron-Cohen, in his essay on the fundamental mechanisms of social cognition, proposes that 'mindreading is simply the best way to make sense of the actions of others' (Baron Cohen, 1997, p. 21) and places ToM at the core of the system of fundamental socio-cognitive mechanisms.

Various tasks have been traditionally used to measure the ToM capabilities in development. In such tasks, participants are usually exposed to descriptions of social situations and are asked to predict the mental states of the characters involved (e.g. the false belief task (Wimmer & Perner, 1983) or the strange stories task (Happé, 1994)). Baron Cohen (1997) proposed the non-verbal Reading the Mind in Eyes test, which measures whether participants can accurately identify internal states from social signals, such as gaze. An alternative widely used test of ToM abilities is the Sally–Anne false belief task (Wimmer & Perner, 1983) in which a cartoon character Sally puts a toy in a basket and leaves the room. Another character, Anne, moves the toy from the basket to a box, while Sally is away. Subsequently, Sally returns to the room. This concludes the sequence of events and participants are asked where they think Sally will look for her toy. The correct answer to this question requires the ability to understand that Sally's knowledge is different from the knowledge of the observer/participant, as Sally was not present in the room when Anne moved the toy. Thus, it requires taking the perspective of Sally and assuming a different mental state (Sally does not know that the toy was moved) than the observer's own mental state (I know that the toy was moved). It also requires the ability to reason about others' mental states. This is why this test has been developed as a test for ToM abilities.

In the context of conceptualisation and operationalisation of ToM, it seems that ToM is another term for the same concept as the one referred to by the term Intentional Stance. While superficially similar indeed, and often equated (also by Baron-Cohen himself), we emphasize the distinction between the concept of the Intentional Stance and ToM. First, ToM—as operationalized in literature—seems to refer to the process of reasoning about another's *specific* mental state, in a specific situation. In the Sally-and-Anne test, Sally has a false belief about the location of her toy.

On the contrary, adoption of the Intentional Stance is a general strategy, which is used to explain and predict a system's behaviour, with reference to mental states. Thus, even though one might attribute a wrong mental state to Sally in the false belief task, they would still be adopting the Intentional Stance towards Sally. Furthermore, the distinction between ToM and the adoption of the Intentional Stance has been supported by the findings of Marchesi et al. (2019), which showed that participants' Intentional Stance scores did not correlate with their scores from the Reading the Mind in the Eyes test (Marchesi et al., 2019).

In the context of a more general landscape of theorising about social cognition mechanisms, Theory of Mind has gained some substantial criticism in literature: Theory of Mind has been proposed as the core of the so-called 'theory theory' (TT), which has been contrasted with the simulation theory (ST) (Apperly & Butterfill, 2009; Butterfill & Apperly, 2013). TT is based on the notion that mental states are not directly accessible to external observers. To understand other individuals, we assume certain mental states and deduce from them predictions regarding the observed behaviour. We then observe the behaviour as 'evidence' for the assumed mental states. Thus, the process of explaining others' behaviour is guided by causal theories that we construct using the information available to us. While all theory-theorists share this assumption, they differ in their views on how the causal theory is generated. Some proponents propose that it is innate, modularized, and develops in tandem with a toddler's cognitive growth, while others argue that it results from bottom-up learning acquired through everyday experiences since birth. Importantly, the theoretical knowledge about others' minds is an automatic process that does not necessitate consciousness (Apperly & Butterfill, 2009).

ST shares the premise with TT that mental states are inaccessible to direct observation. However, it posits that the biological evolution of the human mind implies that mental states are relatively similar across individuals. As a result, there is no need for an organised theory to understand others. Instead, ST suggests that we use our own minds as a model to simulate what 'the minds of others are like'. Different ST theorists have varying views on how this simulation process occurs, either explicitly or implicitly. According to Goldman and Sripada, to understand others, we engage in an explicit process of imagination, simulating what it would be like to be in their shoes (Goldman & Sripada, 2005). However, this approach has faced criticism for being overly conscious and potentially leading to adopting a first-person imaginative perspective.

The discovery of the mirror neuron system in neuroscience has prompted some authors to propose a more implicit version of simulation theory (for a review see Apperly & Butterfill, 2009). According to this perspective, simulation occurs as an enactive process at a sub-personal, implicit, and nearly automatic level. However, one objection raised against this version of ST is that it still appears to imply a first-person perspective. Nevertheless, proponents of this implicit version argue that the activation of the mirror neuron system happens even when an observer merely watches an action being performed by someone else, without physically executing the action themselves. As a result, the simulation is primarily focused on the action itself rather than the specific agent performing it. Thus, the simulation process is described as an enactive, intersubjective mechanism that involves perceiving the action as meaningful, from the perspective of the individuals involved, thereby enabling a deeper understanding of others' actions and intentions (Apperly & Butterfill, 2009).

The ongoing debate between TT and ST regarding social cognition tends to overlook phenomenology. However, certain authors argue for the importance of phenomenology in understanding social cognition, leading to the emergence of three main phenomenological accounts: interaction theory, embodied cognition, and enactive cognition (De Jaegher & Di Paolo, 2007; Fuchs & De Jaegher, 2009). These phenomenological approaches emphasize an embodied and perceptual viewpoint, where social cognition is deeply intertwined with the way we interact with others and the world around us (Gallagher, 2000, 2005; Gallagher & Zahavi, 2020). Thus, social cognition is strongly embedded in the dynamic nature of human interactions.

In sum, there are various theoretical standpoints postulating different views on how people explain the behaviour of others. While the concept of the Theory of Mind—being at the core of the TT account—is perhaps the closest to the concept of the Intentional Stance, the two concepts differ in that ToM is related to the process of reasoning about a specific mental state underlying a specific behaviour while the Intentional Stance is a more general attitude one adopts towards others in explaining and predicting their behaviour.

2.3 Operationalisation of Anthropomorphism and the Intentional Stance

2.3.1 Anthropomorphism as a General Cognitive Bias

Researchers have utilised tools like the Individual Differences in Anthropomorphism Questionnaire (IDAQ) developed by Waytz et al. (2010a, b) to study anthropomorphism. This questionnaire allows for quantifying an individual's tendency towards anthropomorphism, encompassing the attribution of mental states to various non-human entities, including toys, trees, and others. The tool comprises 30 items. Respondents are asked to rate on a 1–10 scale whether they believe a given non-human entity is endowed with certain traits. For example, one of the items asks whether a desert is lethargic, while another item asks whether a river is useful. The questions related to robots ask whether a robot is good-looking or whether a robot has consciousness.

Although the IDAQ questionnaire does not focus specifically on robots, it can provide valuable insights regarding the relationship between individual traits and anthropomorphism. Ruijten and colleagues (Ruijten et al., 2019) have adapted the IDAQ scale to measure specifically anthropomorphism towards robots. They propose the so-called Waytz Instrument which is a 7-item tool with a 5- or 7-point Likert scale rating items such as 'To what extent does the robot have a free will?' or 'To what extend does the robot have desires?'

Marchesi and colleagues (Marchesi et al., 2021), Roselli and colleagues (Roselli et al., 2022a, b), and Spatola et al. (2023) explored how an individual's propensity for anthropomorphism influences their perceptions, attitudes, and interactions with robots. By examining the link between individual differences in anthropomorphism and attitudes towards robots, researchers can gain insights into the psychological factors (at the individual level) that shape Human–Robot Interactions. Such knowledge can guide the development of more personalized and effective human–robot interfaces and facilitate the creation of robots that align better with human expectations and preferences, enhancing the overall user experience. Moreover, understanding the relationship between anthropomorphism and attitudes towards robots can have broader implications for designing human–robot systems that are adaptable, empathetic, and respectful of individual differences in how humans interact with-, and perceive artificial agents.

Similarly to the IDAQ questionnaire, the tool developed by Gray et al. (2007) measures tendencies towards anthropomorphism (mind perception, specifically) in relation to a whole plethora of entities, not limited (but including) robots. In their tool, Gray and colleagues present participants with 78 pairs of schematic drawings of various entities (13 in total). Participants are asked to compare the two entities within each pair on a 5-point scale (e.g. 'which character do you like more' or 'which character is more likely to feel pain'). The characters range from human forms (fetus, infant, adult woman, adult man, or the respondent him/herself) through non-human animals (dog, frog, etc.) to characters such as a dead woman, god, or a robot. The items can be grouped into two main dimensions: capacity for experience (hunger, fear, pain, etc.) and agency (self-control, morality, planning, thought, etc.). The authors have administered this tool to 2399 respondents and have made several interesting observations, for example, that agency is linked to moral responsibility while experience to moral patiency. The authors concluded that mind perception is grounded in those two key distinct dimensions (experience and agency).

2.3.2 *Anthropomorphism Towards Robots*

In the field of HRI, researchers have developed various tools to address attributions of human traits specifically towards robots. These tools are designed to assess specific aspects of the human–robot relationship and evaluate how humans perceive and interact with robots. Some of these tools include:

(a) The Godspeed questionnaire by Bartneck et al. (2009). This tool is perhaps one of the most widely used questionnaires to assess human attitudes towards robots, encompassing the dimensions of anthropomorphism, attribution of animacy, likeability, perceived intelligence, and perceived safety. Twenty-four items of the questionnaire are grouped into these five dimensions. Each item is a 1–5 rating scale placed between two contrasting adjectives, such as ‘anxious’ versus ‘relaxed’ or ‘foolish’ versus ‘sensible’. The questionnaire includes five items related to anthropomorphism, and those are ratings of the following pairs of adjectives: fake versus natural; machine-like versus human-like; conscious versus unconscious; artificial versus lifelike; moving rigidly versus moving elegantly. Respondents are asked to rate their impression of a robot on the scales presented with the 24 items.

(b) Robotic Social Attribute Scale (RoSAS): The RoSAS, developed by Carpinella et al. (Carpinella et al., 2017), has been developed in the context of a critique of the Godspeed questionnaire. More specifically, Carpinella et al. (2017) claimed that several items of the Godspeed scale are confounded with positive and negative affect and that several items do not correspond to the underlying constructs they have been intended to measure. Furthermore, according to Carpinella et al. (2017), four out of the five dimensions of the Godspeed questionnaire are highly correlated, which might indicate that they are measuring similar rather than distinct concepts. As a consequence of this criticism, Carpinella and colleagues proposed a validated scale, Robotic Social Attribute Scale (RoSAS), which consists of modified items of the Godspeed questionnaire complemented with additional items related to more social competencies of a robot, such as friendliness, approachability, and social skills. RoSAS comprises 18 items, where each item is just one adjective and a 9-point Likert scale on which respondents indicate whether a given adjective is associated with the category ‘robots’ (1 = definitely not associated; 9 = definitely associated). The items are grouped into three dimensions: 'warmth', 'competence', and 'discomfort' and include adjectives such as 'compassionate', 'emotional', 'interactive', or 'reliable'.

(c) The Human–Robot Interaction Evaluation Scale (HRIES): The HRIES, introduced by Spatola and colleagues (Spatola et al., 2021), comprises 16 items that are grouped into four dimensions: sociability, agency, animacy, and disturbance. The items consist of an adjective (e.g. 'self-reliant'; 'alive'; 'creepy') and a Likert scale (1–7) to evaluate whether a given adjective corresponds to a robot.

Overall, these tools provide means to evaluate human anthropomorphic attitudes towards robots in general, or towards a specific robot. Oftentimes, however, they are used in relation to respondents' representation of an abstract concept 'robot' and do not capture the impact of various factors (robot appearance, behaviour, or context within which the robot is being presented to participants) on the way humans perceive that specific robot or try to explain/predict its behaviour. In other words, they do not necessarily allow for measuring more nuanced fluctuations of the stance that one might take towards a specific robot in specific circumstances, and how these fluctuations evolve over the course of interaction, or as a function of various experimental manipulations. Furthermore, none of the tools described above address the concept of the Intentional Stance specifically and exclusively.

2.4 Tools to Measure the Adoption of the Intentional Stance

Before moving on to the operationalisation of the concept of the Intentional Stance, let us remind the reader about the concept itself. Dennett (Dennett, 1987) proposes three 'stances' that represent different levels of abstraction for explaining and predicting the behaviour of a system.

1. The Physical Stance focuses on predicting behaviour solely based on the physical properties of the system and the laws of physics. For example, an observer using the physical stance can predict the trajectory of a ball after it is thrown, by considering its initial velocity, angle of release, air resistance, and gravity.
2. The Design Stance involves predicting behaviour based on knowledge of how a system was built to function. For instance, an observer – using the Design Stance – can predict a car's behaviour when the driver pushes the brake pedal by relying on their understanding of the car's functional design.
3. The Intentional Stance considers the mental states and beliefs of the agents to predict their behaviour. For example, an observer adopting the Intentional Stance can predict that a person will order a cup of coffee in a café, by assuming that they have a desire for coffee and that they believe that the café can fulfil that desire.

It is of great interest to understand whether people adopt the Design or the Intentional Stance towards humanoid robots, as robots seem to have the dual nature of being man-made artefacts, on the one hand, and being human-like enough in appearance and behaviour to trigger potentially the adoption of the Intentional Stance. To examine the degree to which people do adopt the Intentional Stance towards robots, one needs to operationalize this high-level philosophical concept.

2.4.1 *The Development of the Intentional Stance (InStance) Test*

Recent literature has made some attempts to operationalize the philosophical concept of the Intentional Stance. For example, Thellman and colleagues conducted a study investigating participants' perception of intentionality in images of humans and humanoid robots, revealing similar levels of intentionality attributed to both (Thellman et al., 2017, see also 2022). The authors presented participants with photographs of a human or a humanoid robot (Pepper) involved in daily activities (mopping the floor; baking a cake). The photographs depicted the human/robot in the context of positively or negatively valenced actions, outcomes, and events of states. The photographs were accompanied by a written description. For example, a positive outcome would be 'Ellis makes fantastic cake' while a negative outcome would be 'Ellis burns the cake'. Participants were asked to rate each scenario (photograph accompanied with a description) on a 7-point Likert scale to what extent the character's behaviour was intentional; under the character's control; or desirable. The results showed that the two characters (human vs. humanoid robot) did not differ much in ratings related to attributed intentionality. However, the limited number of stimuli might have been a limitation regarding the conclusions that one might draw concerning the degree to which humans adopt the Intentional Stance towards robots.

To operationalize the concept of the Intentional Stance, and to overcome the limitations of earlier tools, Marchesi and colleagues developed the InStance Test (IST), a tool to assess individual tendencies to attribute intentionality to humanoid robots. The IST presents participants with 34 pictorial scenarios (cf. Figure 2.1) depicting the humanoid robot iCub (Metta et al., 2010) involved in daily activities

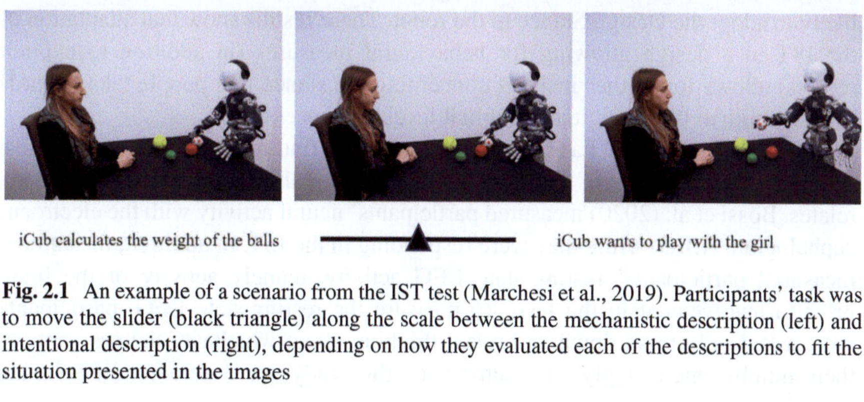

Fig. 2.1 An example of a scenario from the IST test (Marchesi et al., 2019). Participants' task was to move the slider (black triangle) along the scale between the mechanistic description (left) and intentional description (right), depending on how they evaluated each of the descriptions to fit the situation presented in the images

(cleaning the table, playing cards with others, etc.), each scenario accompanied by two descriptions—one mechanistic and one mentalistic. Participants use a slider to indicate the description they believe best fits each scenario (Marchesi et al., 2019). This way, each scenario can receive a 'score', as the position of the slider on the scale can be translated into a numerical value. For example, the scale can be ranging from 0 to 100, where 100 corresponds to the 'intentional' extreme on the scale, while 0 to the 'mechanical' extreme. If a participant is completely convinced that the intentional description of the scenario fits best to what they observe in the scenario, they will move the slider to the 100 extreme. If they are not completely convinced, but they prefer the intentional description to the mechanical one, they might move the slider to the position of about 70% distance between 0 and 100. This way, the test generates a score for every scenario and every participant—a measure of the degree to which Intentional Stance has been adopted towards the iCub humanoid robot. The test has been extensively used across many experiments and samples (Bossi et al., 2020; Marchesi et al., 2019, 2021, 2022) and consistently shows that the baseline IST score is around 40–45, meaning that observers adopt the Intentional Stance towards the humanoid robot to some extent, even though the likelihood to adopt the Design/Mechanical Stance is slightly higher.

In a further development, the IST was adapted to a two-alternative forced-choice task in which the responses to the photos of the iCub from the original IST were compared to photos of a human in the exact same context (Marchesi et al., 2021). Participants were presented with only one of the descriptions (either a mentalistic or a mechanistic description of the scenario), and were asked to respond, as fast as possible, whether they agreed with the description (yes/no). The results showed that when evaluating photos of a human, participants were more likely to accept the mentalistic descriptions relative to mechanistic ones, while in evaluating the robot protagonist, the pattern was reversed. Interestingly, participants were also faster in ascribing the mentalistic descriptions to the human compared to mechanistic ones. For the robot protagonist, there was no difference in speed of responding for these two cases. The authors took this pattern of results to indicate that at the implicit level of processing, participants are equally likely to adopt the Intentional and the Design Stance towards a humanoid robot (no difference in reaction times), even though at the explicit level of processing (acceptance/rejection rates) they are more likely to adopt the Design Stance to the robot. These results show that adaptation of the IST to a design allowing for behavioural measures (in addition to explicit reports) allows for further insights concerning the stance that people take towards robots—both at the explicit and implicit levels of processing.

Interestingly, the IST has been used not only to find a behavioural quantitative measure of the adoption of the Intentional Stance but also to identify its neural correlates. Bossi et al. (2020) measured participants' neural activity with the electroencephalogram (EEG) while they were responding in the IST. In addition, the authors measured participants' resting-state EEG activity, namely activity of the brain prior to involvement in the task. To measure the resting-state EEG, participants were asked only to sit down and relax, thinking about whichever themes came to their minds. Interestingly, the authors of the study were able to differentiate participants' EEG resting-state activity, dependent on whether they were more likely to adopt the Intentional or the Design Stance. More specifically, the sample of participants was split into two groups, based on their average IST score: the group that chose the more 'intentional' responses in the IST and the group that was more likely to choose the more 'mechanistic' responses. Subsequently, the two groups' EEG activity in the resting state was compared and the results showed that the two groups differed with respect to beta activity in the posterior–temporal electrode sides. The authors concluded that the Intentional Stance can be correlated with a specific pattern of resting-state EEG activity. Interestingly, this pattern of activity has been observed in literature as a marker for the so-called default-mode network (Laufs et al., 2011)—a network observed during the resting state (thus, when the participants are not engaged in any specific experimental task), which also largely overlaps with the network of the so-called 'social brain' (Meyer, 2019)—a network of neural regions activated in social cognition tasks.

The study of Bossi et al. (2020) was, however, not the first to identify neural correlates of the Intentional Stance. The seminal study of Gallagher et al. (2002) identified patterns of brain activity specifically associated with the adoption of the Intentional Stance. The authors scanned participants' brain activity using positron emission tomography (PET), while they were performing a computerized version of the rock-paper-scissors game. Then, the experimenters asked participants to second-guess the opponent. In one condition, participants believed that they were playing against the experimenter. In the other condition, they believed they were playing against a computer. However, in both conditions, the 'opponent' produced a random sequence, and thus the only difference was the attitude (or stance) adopted by participants throughout the task. By analysing participants' brain activity, the authors showed that the anterior paracingulate cortex showed differential activation, dependent on the stance participants took. Interestingly, this brain area is part of the default mode network (Amodio & Frith, 2006) and is also involved in mentalizing (Mars et al., 2012; Bolis & Schilbach, 2018).

Furthermore, Chaminade et al. (2012) scanned participants using fMRI during interactive rock-paper-scissors games, while they believed that their opponent was a fellow human, a humanoid robot endowed with artificial intelligence, or a computer playing randomly. Interestingly, the authors found that the mentalizing regions (mPFC, TPJ) responded only when the opponent was another human, supporting the specificity of mentalizing areas only for interactions with clearly intentional agents. In other words, Chaminade et al.' (2012) results suggest that people did not involve mentalizing areas when observing robots.

Finally, Cohen et al. (2009) identified gamma-band activity in the superior temporal sulcus (STS), a cortical area that is part of the 'social brain', as a marker of mentalization for non-verbal social cues (Cohen et al., 2009). In the study of Cohen et al. (2009), participants were asked to interpret the desires and intentions of a virtual character based on the combination of different facial expressions, gestures, and body postures, and they found that the gamma-band activity correlated with the processes necessary to complete this task.

2.5 Conclusions

This chapter delineated the concept of the Intentional Stance and highlighted that this concept is closely related, but not identical, to concepts such as anthropomorphism or Theory of Mind. Furthermore, by reviewing empirical studies related to anthropomorphism, Intentional Stance, and Theory of Mind, the chapter illustrated that the concepts differ not only at the theoretical, but also at the empirical level. The chapter's last part described in detail measures of Intentional Stance both at the behavioural and at the neural level, and focused on the tool that has been developed to operationalize the Intentional Stance in relation to humanoid robots (the Intentional Stance Test by Marchesi et al., 2019). This test has been used extensively in a variety of studies investigating factors that influence the likelihood of adopting the Intentional Stance, described in the following chapters.

References

Airenti, G. (2018). The development of anthropomorphism in interaction: Intersubjectivity, imagination, and theory of mind. *Frontiers in Psychology, 9*, 2136.

Airenti, G., Cruciani, M., & Plebe, A. (2019). Editorial: The cognitive underpinnings of anthropomorphism. *Frontiers in Psychology, 10*, 1539. https://doi.org/10.3389/fpsyg.2019.01539

Amodio, D. M., & Frith, C. D. (2006). Meeting of minds: The medial frontal cortex and social cognition. *Nature Review Neuroscience, 7*, 268–277. https://doi.org/10.1038/nrn1884

Andrews, K., & Huss, B. (2014). Anthropomorphism, anthropectomy, and the null hypothesis. *Biology & Philosophy, 29*(5), 711–729. https://doi.org/10.1007/s10539-014-9442-2

Apperly, I. A., & Butterfill, S. A. (2009). Do humans have two systems to track beliefs and belief-like states? *Psychological Review, 116*(4), 953–970. https://doi.org/10.1037/a0016923

Baron-Cohen. (1997). *Mindblindness. . An essay on autism and theory of mind.* MIT Press. First MIT paperback edition.

Barrouillet, P. (2015). Theories of cognitive development: From piaget to today. *Developmental Review, 38*, 1–12. https://doi.org/10.1016/j.dr.2015.07.004

Bartneck, C., Kulić, D., Croft, E., & Zoghbi, S. (2009). Measurement instruments for the anthropomorphism, animacy, likeability, perceived intelligence, and perceived safety of robots. *International Journal of Social Robotics, 1*, 71–81.

Bolis, D., & Schilbach, L. (2018). Observing and participating in social interactions: Action perception and action control across the autistic spectrum. *Developmental Cognitive Neuroscience, 29*, 168–175.

Bossi, F., Willemse, C., Cavazza, J., Marchesi, S., Murino, V., & Wykowska, A. (2020). The human brain reveals resting state activity patterns that are predictive of biases in attitudes toward robots. *Science Robotics, 5*(46), eabb6652. https://doi.org/10.1126/scirobotics.abb6652

Butterfill, S. A., & Apperly, I. A. (2013). How to construct a minimal theory of mind: How to construct a minimal theory of mind. *Mind & Language, 28*(5), 606–637. https://doi.org/10.1111/mila.12036

Caporael, L. R., & Heyes, C. M. (1997). Why anthropomorphize? Folk psychology and other stories. *Anthropomorphism, Anecdotes, and Animals*, 59–73.

Carpinella, C. M., Wyman, A. B., Perez, M. A., & Stroessner, S. J. (2017). *The robotic social attributes scale (RoSAS): Development and validation.* In Proceedings of the 2017 ACM/

IEEE international conference on human-robot interaction (pp. 254–262). https://doi.org/10.1145/2909824.3020208.

Chaminade, T., Rosset, D., Da Fonseca, D., Nazarian, B., Lutcher, E., Cheng, G., & Deruelle, C. (2012). How do we think machines think? An fMRI study of alleged competition with an artificial intelligence. *Frontiers in Human Neuroscience, 6*, 103. https://doi.org/10.3389/fnhum.2012.00103

Cohen, M. X., David, N., Vogeley, K., & Elger, C. E. (2009). Gamma-band activity in the human superior temporal sulcus during mentalizing from nonverbal social cues. *Psychophysiology, 46*(1), 43–51.

Dacey, M. (2017). Anthropomorphism as cognitive bias. *Philosophy of Science, 84*(5), 1152–1164. https://doi.org/10.1086/694039

De Jaegher, H., & Di Paolo, E. (2007). Participatory sense-making: An enactive approach to social cognition. *Phenomenology and the Cognitive Sciences, 6*(4), 485–507. https://doi.org/10.1007/s11097-007-9076-9

Dennett, D. C. (1987). *The intentional stance*. MIT Press.

Epley, N., Waytz, A., & Cacioppo, J. T. (2007). On seeing human: A three-factor theory of anthropomorphism. *Psychological Review, 114*(4), 864–886. https://doi.org/10.1037/0033-295X.114.4.864

Fisher, J. A. (1991). Disambiguating anthropomorphism: An interdisciplinary review. *Perspectives in Ethology, 9*(9), 49–85.

Frith, C. D., & Frith, U. (2006). The neural basis of mentalizing. *Neuron, 50*(4), 531–534.

Fuchs, T., & De Jaegher, H. (2009). Enactive intersubjectivity: Participatory sense-making and mutual incorporation. *Phenomenology and the Cognitive Sciences, 8*(4), 465–486. https://doi.org/10.1007/s11097-009-9136-4

Gallagher, S. (2000). Philosophical conceptions of the self: Implications for cog-nitive science. *Trends in Cognitive Sciences, 4*(1), 8.

Gallagher, S. (2005). *How the body shapes the mind*. Oxford University Press. https://doi.org/10.1093/0199271941.001.0001

Gallagher, S., & Zahavi, D. (2020). *The phenomenological mind: An introduction to philosophy of mind and cognitive science* (3rd ed.). Routledge.

Gallagher, H. L., Jack, A. I., Roepstorff, A., & Frith, C. D. (2002). Imaging the intentional stance in a competitive game. *NeuroImage, 16*(3), 814–821. https://doi.org/10.1006/nimg.2002.1117

Goldman, A. I., & Sripada, C. S. (2005). Simulationist models of face-based emotion recognition. *Cognition, 94*(3), 193–213. https://doi.org/10.1016/j.cognition.2004.01.005

Gray, H. M., Gray, K., & Wegner, D. M. (2007). Dimensions of mind perception. *Science, 315*(5812), 619–619.

Guthrie, S. E. (1997). Anthropomorphism: A definition and a theory. In R. W. Mitchell, N. S. Thompson, & H. L. Miles (Eds.), *Anthropomorphism, anecdotes, and animals* (pp. 50–58). State University of New York Press.

Happé, F. G. (1994). An advanced test of theory of mind: Understanding of story characters' thoughts and feelings by able autistic, mentally handicapped, and normal children and adults. *Journal of Autism and Developmental Disorders, 24*(2), 129–154.

Heider, F. (1958). Perceiving the other person.

Heider, F., & Simmel, M. (1944). An experimental study of apparent behavior. *The American Journal of Psychology, 57*(2), 243–259.

Jones, R. A. (2022). Projective anthropomorphism as a dialogue with ourselves. *International Journal of Social Robotics, 14*(10), 2063–2069. https://doi.org/10.1007/s12369-021-00793-7

Kahn, P. H., & Shen, S. (2017). NOC NOC, who's there? A new ontological category (NOC) for social robots. In N. Budwig, E. Turiel, & P. D. Zelazo (Eds.), *New perspectives on human development* (1st ed., pp. 106–122). Cambridge University Press. https://doi.org/10.1017/CBO9781316282755.008

Laufs, H., Krakow, K., Sterzer, P., Eger, E., Beyerle, A., Salek-Haddadi, A., & Kleinschmidt, A. (2011). Electroencephalographic signatures of attentional and cognitive default modes in spontaneous brain activity fluctuations at rest. *Proceedings of the National Academy of Sciences United States of America, 100*, 11053–11058.

Marchesi, S., Ghiglino, D., Ciardo, F., Perez-Osorio, J., Baykara, E., & Wykow-ska, A. (2019). Do we adopt the intentional stance toward humanoid robots? *Frontiers in Psychology, 10*, 450. https://doi.org/10.3389/fpsyg.2019.00450

Marchesi, S., Spatola, N., Perez-Osorio, J., & Wykowska, A. (2021). *Human vs humanoid. A Behavioral investigation of the individual tendency to adopt the intentional stance*. In HRI '21: Proceedings of the 2021 ACM/IEEE international conference on human-robot interaction (pp. 332–340). https://doi.org/10.1145/3434073.3444663.

Marchesi, S., De Tommaso, D., Perez-Osorio, J., & Wykowska, A. (2022). Belief in sharing the same phenomenological experience increases the likelihood of adopting the intentional stance toward a humanoid robot. 11.

Mars, R. B., Neubert, F. X., Noonan, M. P., Sallet, J., Toni, I., & Rushworth, M. F. (2012). On the relationship between the "default mode network" and the "social brain". *Frontiers in Human Neuroscience, 6*, 189.

Metta, G., Natale, L., Nori, F., Sandini, G., Vernon, D., Fadiga, L., von Hofsten, C., Rosander, K., Lopes, M., Santos-Victor, D., Bernardino, A., & Montesano, L. (2010). The iCub humanoid robot: An open-systems platform for research in cognitive development. *Neural Networks, 23*(8–9), 1125–1134.

Meyer, M. L. (2019). Social by default: Characterizing the social functions of the resting brain. *Current Directions in Psychological Science, 28*, 380–386.

Mithen, S., & Boyer, P. (1996). Anthropomorphism and the evolution of cognition. *Journal of the Royal Anthropological Institute*, 717–721.

Piaget, J. (1929). Les deux directions de la pensée scientifique. *Archive des Sciences Physiques et Naturelles, 11*, 145–162.

Premack, D., & Woodruff, G. (1978). Does the chimpanzee have a theory of mind? *Behavioral and Brain Sciences, 1*(4), 515–526.

Roselli, C., Ciardo, F., De Tommaso, D., & Wykowska, A. (2022a). Human-likeness and attribution of intentionality predict vicarious sense of agency over humanoid robot actions. *Scientific Reports, 12*(1), 13845. https://doi.org/10.1038/s41598-022-18151-6

Roselli, C., Ciardo, F., & Wykowska, A. (2022b). Intentions with actions: The role of intentionality attribution on the vicarious sense of agency in human–robot interaction. *Quarterly Journal of Experimental Psychology*.

Ruijten, P. A. M., Haans, A., Ham, J., & Midden, C. J. H. (2019). Perceived human-likeness of social robots: Testing the Rasch model as a method for measuring anthropomorphism. *International Journal of Social Robotics, 11*(3), 477–494.

Spatola, N., Kühnlenz, B., & Cheng, G. (2021). Perception and evaluation in human–robot interaction: The human–robot interaction evaluation scale (HRIES)—A multicomponent approach of anthropomorphism. *International Journal of Social Robotics, 13*(7), 1517–1539. https://doi.org/10.1007/s12369-020-00667-4

Spatola, N., Marchesi, S., & Wykowska, A. (2022). Different models of anthro-pomorphism across cultures and ontological limits in current frameworks the integrative framework of anthropomorphism. *Frontiers in Robotics and AI, 16*.

Spatola, N., Marchesi, S., & Wykowska, A. (2023). The phenotypes of anthro-pomorphism and the link to personality traits. *International Journal of Social Robotics, 15*(1), 3–14. https://doi.org/10.1007/s12369-022-00939-1

Surden, H., & Williams, K. C. (2016). Technological opacity, predictability, and self-driving cars. *19*(2), 165–198.

Thellman, S., Silvervarg, A., & Ziemke, T. (2017). Folk-psychological interpretation of human vs. humanoid robot behavior: Exploring the intentional stance toward robots. *Frontiers in Psychology, 8*, 1962. https://doi.org/10.3389/fpsyg.2017.01962

Thellman, S., de Graaf, M., & Ziemke, T. (2022). Mental state attribution to robots: A systematic review of conceptions, methods, and findings. *ACM Transactions on Human-Robot Interaction, 3526112*. https://doi.org/10.1145/3526112

Urquiza-Haas, E. G., & Kotrschal, K. (2015). The mind behind anthropomorphic thinking: Attribution of mental states to other species. *Animal Behaviour, 109*, 167–176. https://doi.org/10.1016/j.anbehav.2015.08.011

Waytz, A., Cacioppo, J., & Epley, N. (2010a). Who sees human?: The stability and importance of individual differences in anthropomorphism. *Perspectives on Psychological Science, 5*(3), 219–232. https://doi.org/10.1177/1745691610369336

Waytz, A., Epley, N., & Cacioppo, J. T. (2010b). Social cognition unbound: In-sights into anthropomorphism and dehumanization. *Current Directions in Psychological Science, 19*(1), 58–62. https://doi.org/10.1177/0963721409359302

Wimmer, H., & Perner, J. (1983). Beliefs about beliefs: Representation and constraining function of wrong beliefs in young children's understanding of deception. *Cognition, 13*(1), 103–128.

Chapter 3
Factors Influencing the Adoption of the Intentional Stance: The Human Observer

Cecilia Roselli, Serena Marchesi, Nicolas Spatola, Davide Ghiglino, Uma Prashant Navare, and Francesco Bossi

We cannot safely assume that other people's minds work on the same principles as our own.
All too often, others with whom we come in contact do not reason as we reason, or do not value the things we value, or are not interested in what interests us.

– Isabel Briggs Myers (1897–1980).

3.1 Introduction

In everyday life, we usually appreciate the (substantial) differences among the people we are surrounded by. We do so because such differences introduce a rich diversity of cultures, perspectives, and lifestyles into our societies. However, in the neuroscience of human behaviour and cognition, interindividual differences are often treated as a source of 'noise', and thus discarded or neglected. For example, a common practice in experimental psychology is to average data from a group of

C. Roselli (✉) · S. Marchesi · D. Ghiglino
Italian Institute of Technology, Social Cognition in Human-Robot Interaction Unit, Genova, Italy
e-mail: cecilia.roselli@unimib.it

N. Spatola
Artimon Perspectives, Paris, France

U. P. Navare
Italian Institute of Technology, Social Cognition in Human-Robot Interaction Unit, Genova, Italy

CY Cergy Paris Université, ENSEA, CNRS, Cergy, France

F. Bossi
Department of Information Engineering, University of Pisa, Pisa, Italy

A. Wykowska (ed.), *Intentional Stance Towards Humanoid Robots*,
https://doi.org/10.1007/978-3-031-65483-1_3

participants, which might lead to conclusions that ignore a substantial difference in individual responses. Another common practice is, when recruiting a sample of participants for an experiment, to restrict only to the W.E.I.R.D. (Western, Educated, Industrialized, Rich, and Democratic) individuals. This is presumably because this community is usually the most accessible. However, one needs to keep in mind that they constitute only 12% of the world population (Henrich et al., 2010). In effect, findings from studies involving this restricted sample of individuals are often generalized to the world population—both in terms of socio-cognitive processes, such as decision-making, auditory and visual perception, as well as psychological traits (Henrich et al., 2010).

This said, it is crucial to recognize that human cognition might not always operate the same way for all individuals. For instance, potential sources of variation in how people approach cognitive tasks might include individual differences in cognitive abilities and cognitive styles, as well as in other characteristics such as sex, gender, and age. These differences become particularly relevant in the context of social cognition, namely the set of high-level cognitive processes subserving the extremely diverse and flexible social behaviours that we spontaneously adopt when interacting with other conspecifics (Adolphs, 1999). Interindividual differences are relevant because they can substantially change the way we approach others, and in consequence, their responses to our behaviour. Thus, they definitely shape our interactions with others. This is the reason why elucidating the contribution of individual differences is central to our understanding of social behaviour (Garfinkel et al., 2015).

The same concern regarding the relevance of individual differences applies also to interaction with non-human and artificial agents, such as robots. Fischer (2011) argued that there is a considerable amount of interpersonal variation with respect to whether artificial agents, such as robots, are treated as social actors. One important aspect in the context of social cognition in Human–Robot Interaction (HRI) is the target population that robots are designed for, and the purpose of the interaction (Henschel et al., 2020). Indeed, robots designed for elderly assistance, or as supportive technologies in children's educational settings, should satisfy different requirements, and thus, individuals' characteristics need to be considered when developing the design, inner functioning, and usability of social robots.

Luckily, in the past decades, the HRI community has gradually extended its focus to broader geographical and societal representativeness, and addressed differences in people's expectations, attitudes, and beliefs towards robots. For example, Epley and colleagues (2007) suggested that people's likelihood of anthropomorphizing a mechanical agent (e.g. a robot) can be predicted by interpersonal differences in three determinants—the amount of knowledge about the other agent, the motivation to understand their behaviour, and the desire of social contact and affiliation (Epley et al., 2007).

Several studies have been conducted to explore the relationship between individual traits and anthropomorphism, especially in the context of Human–Robot Interaction. For example, researchers explored personality traits (Spatola et al., 2023), spiritualistic tendencies and cultural differences (Spatola et al., 2022a; Tanyi, 2002), need for cognition and need for closure (Spatola & Wykowska, 2021) to identify the potential impact of such tendencies on anthropomorphism (Spatola et al., 2022b). The results highlighted a strong relationship between these constructs in such a way that researchers identified clusters of individuals sharing certain personality traits. This chapter will describe research addressing the question of whether, and how, interindividual differences are at play in modulating people's perception of robots. Our special focus is the Intentional Stance—in other words, whether, and how, individual differences modulate people's tendency to explain and predict behaviour of robots with reference to their (robots') mental states.

3.2 Individual Differences and Anthropomorphism

To effectively navigate the uncertain and ambiguous external world, individuals employ strategies to make sense of their environment (Bargh et al., 2001; Gollwitzer & Sheeran, 2006). Anthropomorphism and Intentional Stance serve as intuitive and accessible strategies to reduce complexity and uncertainty. This is particularly relevant when individuals encounter non-human agents for which they lack alternative models allowing to explain their behaviour (Maner et al., 2005). The motivation to resolve uncertainty, seek meaning, and feel efficacious influences the degree of anthropomorphism and intentional attributions. Epley et al. (2007) suggest that attributing human characteristics to non-human agents may enhance understanding and predictability, while also providing a sense of control over the situation (need for closure) (Epley et al., 2008).

According to Barsalou (1983), Epley et al. (2007), Spatola and Wykowska (2021), and Urquiza-Haas and Kotrschal (2015), attributing human cognitive capacities to non-human entities can be seen as a process of inductive inference, where individuals use available information to interpret an observed entity within a framework of accessible representation. People have direct access to their own phenomenological experience of being human, allowing them to understand and infer mental states of others based on their own. However, representing the experience of being a non-human agent, such as a robot, is cognitively demanding. The extent to which someone adopts the Intentional Stance in the face of a non-human agent depends on cognitive factors that moderate the weight of external (coming from the observed agent) and internal (coming from the observer) information. This interplay between the natural tendency to use internal knowledge and cognitive control is influenced by individuals' motivation and capability to understand and predict others, known as the *need for cognition* (Cacioppo et al., 1984). Individuals with a higher need for cognition might be more open to new information and alternative (to the readily available) models of explanations (Cacioppo et al., 1996; Epley et al., 2007), which might result in a lower likelihood of anthropomorphizing (Spatola & Chaminade, 2022; Spatola & Wykowska, 2021).

The concept of the *need for closure*, proposed by Kruglanski (1990) and Webster and Kruglanski (1994), provides a theoretical framework for understanding the cognitive and motivational aspects of decision-making, which can vary among individuals. The five dimensions of the need for closure include the preference for order, predictability, decisiveness, discomfort towards ambiguity, and close-mindedness (Webster & Kruglanski, 1994). Individuals with a high need for closure and a desire for control tend to rely on easily accessible information rather than investing effort in acquiring alternative models, new knowledge, or constructing more accurate representations (Epley et al., 2008). Humans still find it unusual to interact with artificial agents; there is inherent difficulty in comprehending their capabilities, leading to ambivalent attitudes and feelings towards robots (Sundar et al., 2016). Therefore, anthropomorphism and Intentional Stance become convenient solutions when encountering robots, as attributing human characteristics makes the situation more comprehensible. Indeed, results of Spatola and Wykowska (2021) showed that both personality dimensions—need for cognition and need for closure—play a role in shaping attitudes towards robots and anthropomorphic attributions, and this effect is further emphasized by the human-like characteristics of the robots, unravelling the importance of considering the interplay between dispositional traits and robot design in Human–Robot Interaction.

3.3 Individual Differences and Adoption of the Intentional Stance

When validating the InStance Test (IST), Marchesi and colleagues (2019) observed individual differences in people's likelihood of adopting the Intentional Stance towards robots. The authors observed that a subsample of participants was more prone to choose mentalistic explanations of robots' behaviours than another subsample, which was more disposed towards mechanistic explanations, that is, to choose sentences describing iCub's behaviour with reference to its mechanical functioning (Marchesi et al., 2019).

Bearing in mind the differences between the two subsamples of participants, observed in the IST validation, and an analogy to anthropomorphism, researchers have addressed interindividual differences in the likelihood of adopting the Intentional Stance towards robots. This paragraph will focus on two interindividual determinants that might be at play in the context of adoption of the Intentional Stance:

- Individuals' state of mind before or during the interaction.
- Prior knowledge or expertise regarding robots.

3.3.1 State of Mind

Addressing the 'state of mind' of participants when they are just about to engage in an interaction with a robot is perhaps best achieved by examining the so-called default mode network (DMN). DMN is a large-scale brain network that is active during periods of rest, namely when the mind is not focused on any specific task.

Conversely, its activity decreases when individuals' attention is oriented towards a task in an external environment. Thus, the prevailing theory regarding the function of the DMN suggests its involvement in self-referential processing (Buckner et al., 2008).

Notably, patterns of activation of the DMN seem to be very similar to those observed in various aspects of social cognition, including the involvement of the precuneus in social interactions (Schilbach et al., 2006), the left angular gyrus/temporoparietal junction (TPJ) in differentiating between self and others (Vogeley & Fink, 2003), and anterior cingulate cortex (ACC) in action monitoring for self and others (Amodio & Frith, 2006). In addition, previous evidence has revealed a well-established anatomical overlap between the DMN and the cortical areas involved in mentalizing (Meyer, 2019) or adoption of the Intentional Stance (Spunt et al., 2015).

A study by Spunt et al. (2015) showed that the DMN primes the adoption of the Intentional Stance. In other words, activity in the DMN during periods of rest (and before encountering a social stimulus) might serve as an endogenous prime that makes it more likely to adopt the Intentional Stance, thus using it as a default strategy for making sense of the social world (Spunt et al., 2015). Spunt and colleagues' setup was somewhat comparable to the IST (Marchesi et al., 2019), but with a key difference—the former pertains to explanations of human behaviours, whereas the latter centres on attributing intentionality to robots. Spunt et al. (2015) discussed how the Intentional Stance involves self-referential processes more than goal-directed cognition, even though the attribution is oriented towards external agents. Thus, it appears that the DMN's self-referential processes and the attribution of intentions to another agent are strictly connected in our brains.

Since the DMN comprises several neural areas in a very articulated network, its investigation is often approached using neuroimaging techniques, such as functional magnetic resonance imaging (fMRI). Indeed, this family of research techniques is extremely accurate in identifying the spatial location of specific areas being active in the brain thanks to the detection of the so-called blood-oxygen-level-dependent (BOLD) signal (Gazzaniga, 2004). In other words, the fMRI can identify which areas are active by measuring the flow of oxygenated blood, which is necessary for brain cell metabolism. Unfortunately, this kind of technique also presents low temporal definition as the blood flow takes a few seconds to reach the activated area. Another technique often used in neuroscience that can obtain much higher temporal resolution is electroencephalography (EEG). EEG records real-time electrical activity of the brain, given that neurons communicate with each other via electrochemical signals. The downside of this technique is lower spatial resolution due to the inverse problem of source localization of the EEG signal recorded on the surface of the scalp. However, a neural signal recorded with such temporal definition can allow for different types of analyses, such as time–frequency representations (TFR) of the signal to study oscillatory components of the EEG signal. Typically, explored frequencies range between 1 Hz (delta range, associated with deep sleep) and 100 Hz (gamma range, high-frequency oscillatory activity). Different frequency bands can represent different and specific cognitive processes involved in a given task.

Concerning the cognitive processes of our interest, beta-band oscillations (13–30 Hz) are a consistent indicator of spontaneous cognitive processes during conscious rest, and they exhibit a robust correlation with the activation of cortical areas engaged in DMN (Mantini et al., 2007), particularly the medial and lateral parietal regions (Jann et al., 2010).

Bossi et al. (2020) investigated whether a specific pattern of EEG activity at rest could be related to individual attitudes towards robots. The authors asked participants to complete the InStance Test (IST; Marchesi et al., 2019) while recording their brain activity via EEG. Most importantly, they recorded individuals' resting-state EEG activity, before they engaged in the IST, and were still naïve to the purpose of the study. Then, neural activity of participants who were more likely to provide an intentional explanation of iCub's behaviours in the IST was compared to the neural activity of those who were more likely to provide mechanical (Design Stance) explanations.

Participants who were biased towards attributing a mechanistic explanation to iCub's behaviour (i.e. the Design Stance) exhibited increased bilateral parietal activity in the beta range, compared to those who were more likely to adopt the Intentional Stance towards the robot. Notably, this pattern of activity during the resting state had been associated with the DMN (Mantini et al., 2007); therefore, the authors concluded that participants adopting the Design Stance towards a robot tend to show an increased use of the DMN at rest. This finding might at first appear as counterintuitive, as previous evidence showed that the DMN activation primes the adoption of the Intentional Stance when participants provided explanations for humans' behaviours portrayed in pictures (Spunt et al., 2015). In this context, 'human' might be the key word to interpret this result. Indeed, the findings of Bossi et al. (2020) showed that people who, during resting state, were activating to a higher degree the DMN (thus, were likely thinking about social situations, typically involving other humans), were less likely to use mentalistic explanations in relation to a robot. This might be because after having thought about other humans and their mental states, seeing a robot during the IST contrasted more with their mindset during the earlier resting state, in comparison to participants who have not been thinking much about social situations (and thus activated the DMN to a lesser extent) during the resting state.

Building on these results, Marchesi and colleagues further explored the impact of individuals' state of mind on the propensity to adopt the Intentional Stance towards robots, using another interesting measure: pupil dilation (Marchesi et al., 2021a). Similarly to the study of Bossi et al. (2020), Marchesi et al. (2021a) observed different patterns of pupil dilation as a function of individual tendencies to adopt the Intentional Stance or the Design Stance. Interestingly, the authors found that also RTs were predictive of participants' individual biases in adopting the Intentional Stance versus the Design Stance (Marchesi et al., 2021a).

Overall, the results described here suggest that a given state of mind—as indicated by neural, physiological, and behavioural markers—might influence one's propensity to adopt the Intentional Stance towards a humanoid robot. This has not only theoretical, but also practical, implications: it might allow, in the future, to design and develop robots that will be able to adapt their behaviour to the specific state of mind of a user at a given time of interaction.

3.3.2 Prior Knowledge and Expertise

Independent of whether we adopt the Design or the Intentional Stance towards others, the type of knowledge we put at play (regarding the way we understand and predict others' behaviour) is what guides us through the interaction, and thus can deeply affect our experience of it (Emberson, 2016; Hudson et al., 2016; Laurent et al., 2015). When considering the interaction between humans and robots, the type of knowledge (prior knowledge, expertise) that we have might change our perception of robots, namely whether we perceive them as mechanical artefacts (adopting the Design Stance), or as agents endowed with mental states (adopting the Intentional Stance).

In the next sections, we will use the following terminology, partially inspired by the American Psychological Association's (APA) *Dictionary of Psychology* (https://dictionary.apa.org/). '*Prior knowledge*' will denote knowledge consolidated before a given time (e.g. before the beginning of the experimental session) and being acquired through repetitive interactions or dedicated training. '*Expertise*' will denote a high level of domain-specific knowledge accumulated over time, usually through long-term cumulative experiences (e.g. a formal education).

3.3.2.1 Prior Knowledge

One interesting question in the context of the Intentional Stance topic is the extent to which its usefulness, and predictive power, can be modulated by providing information about the capabilities of the observed agents before the interaction with them (Thellman & Ziemke, 2019). In Human-Human Interaction (HHI), individuals usually go through a fast knowledge estimation process in which they exchange information about goals and intentions with the partner(s). In this context, each agent develops knowledge about the other (Isaacs & Clark, 1987). To this aim, people use a series of observable, physical (e.g. motion), and behavioural cues (e.g. non-verbal movements, actions, or gestures), as well as contextual information which might enrich and provide meaningful interpretations of these cues (Fussel & Krauss, 1992). For instance, information about the other's country of origin, group membership, educational background, and professional identity.

When the other agent is a robot, the process is similar; individuals still build a certain knowledge by relying on a series of cues as it is the case with other humans (Kiesler, 2005). Therefore, people's knowledge is sensitive to the physical attributes of the robot (Powers & Kiesler, 2006), as well as to the information about the robot acquired before the interaction, such as its inner capabilities (Lee et al., 2005; Powers & Kiesler, 2006). This said, a question arises naturally: Would repeated interactions with a robot, in such a way that individuals can form their own knowledge about the robot, help individuals to treat them as agents endowed with mental states, that is, eliciting the adoption of the Intentional Stance towards them?

Past research provided some hints: repeated interactions with a robot tend to enhance the perception of the robot as a socially present agent, and thus, as a social partner (Fiore et al., 2013; Häring et al., 2014; Deligianis et al., 2017). Already in

1954, Allport and colleagues proposed an influential theory called the Contact Hypothesis, according to which interpersonal contacts among members of different social groups significantly contribute to reducing, over time, negative attitudes such as prejudice and conflicts (Allport et al., 1954). In other words, repeated interactions allow individuals to acquire a rich and well-grounded knowledge about the other agent, which increases the perceived similarity between 'me' and the 'other'. As a result, people would be less prone to exhibit negative attitudes towards the 'other'. The assumption behind the Contact Hypothesis—that is, the idea that repeated interactions promote reciprocal positive attitudes among inter-group members—might be applied also to the context of HRI. Specifically, a context in which the 'other' is not simply a human member of another social group, but a robot—something that, by nature, is further away from us than any other human being will ever be. Therefore, when repeatedly engaging in a task with a robot, humans might start considering it over time as a member of their own in-group, showing the same positive attitudes that they usually display with another in-group member.

Ciardo and colleagues (2020) evaluated individual differences in attitudes towards robots, before and after repeated interactions with them. More specifically, the authors examined individual attitudes towards robots in a group of clinicians who took part in a robot-assisted training for children with autism spectrum disorders over several weeks—thus, the clinicians were exposed to repeated interactions with a non-anthropomorphic robot (the Cozmo robot by Anki Robotics). The clinicians were tested at three different time points: at the beginning of the training, at the middle, and at the end of the training. Results showed that the more clinicians were exposed to the robot during training with children, the more they tended to perceive the robot as a social partner (Ciardo et al., 2020). This evidence suggests that indeed repeated interaction with robots makes people perceive them more as social partners, and less as purely mechanical artefacts, which would be in line with results from social psychology about human–human interactions.

However, in another study by Roselli and colleagues (Roselli et al., 2024), the authors examined social inclusion (as in the case of Ciardo et al., 2020), as a function of structured training. In this study, participants were engaged in the embodied version of the Cyberball game (Williams & Jarvis, 2006), before and after the same training that Roselli et al. (2024) administered. Participants performed the game in a physical space together with another human player—the confederate of the experimenter—and iCub (see Fig. 3.1).

Interestingly, participants tended to perceive iCub as a social partner (by including the robot in the game) in both sessions of the Cyberball, regardless of the technical knowledge they acquired during the technical training phase. In other words, the fact that participants were exposed to training, which was supposed to provide them with a formal knowledge about the mechanical functioning of the robot, did not modulate their social inclusion of the robot.

Both results were somewhat surprising, as one would expect that a formal training about robots' mechanisms and mechanical functioning should have an impact on the perception of robots as social partners and/or on the adoption of the Intentional Stance.

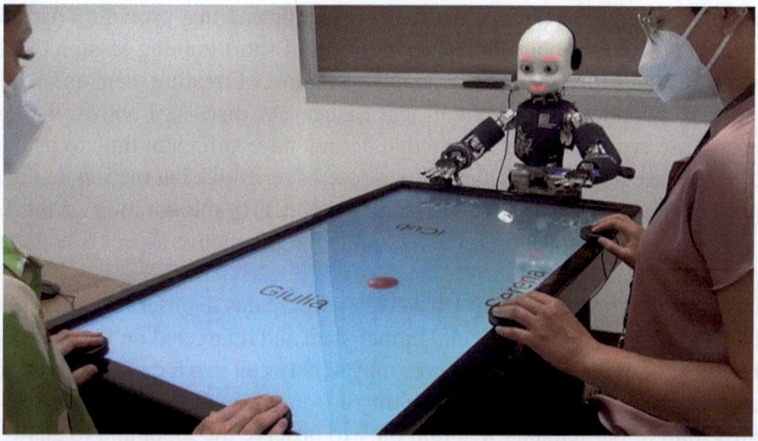

Fig. 3.1 Embodied version of the original Cyberball paradigm (Williams & Jarvis, 2006), in which participants performed the ball game with another human confederate and the fully embodied, humanoid robot iCub. The picture was taken from the original paper by Roselli et al. (2024), with the permission of all the authors

To interpret these unexpected results, there are two important aspects to be considered: (i) the duration, and (ii) the format of the training. A structured training in a form of a 'lecture' does not affect individuals' attitudes towards the robot. Perhaps what is needed is the first-hand experience of interaction, as in the case of the results reported by Ciardo et al. (2020), where the participants (clinicians) interacted with the robot over an extended period of time, while providing therapeutic sessions to children. The second aspect is time (duration)—indeed, a short presentation of the 'mechanical workings' of the robot might have not allowed for consolidation of this knowledge and sufficiently deep processing of this information. In consequence, it had no impact on the social inclusion of robots as in-group members.

One important question to be asked in the context of these results, which involved adult participants, is whether these findings are generalizable to other populations, such as children.

Following this line of reasoning, O'Reilly et al. (under review) performed a study in which middle school children were involved in a 2-h hands-on workshop, focusing on programming robotic components. Children were first shown videos in which the humanoid robot iCub displayed human-like behaviours (e.g. doing math calculations, or performing tai-chi), and then a 3D-printed humanoid robot arm moving, while the experimenter explained the technical properties of iCub to clarify the technology underlying the human-like behaviours of the robot. Both before and after the technical workshop, children's likelihood of adopting the Intentional Stance was assessed by means of the IST (Marchesi et al., 2019). The results showed that participating in the technical workshop did not significantly change children's tendency to attribute intentional traits to robots (i.e. IST scores). This paralleled the results obtained with adults, as described above.

Taken together, the evidence outlined so far showed that providing participants with prior knowledge about robots, in the form of short training sessions, does not affect the attitudes towards robots—neither in terms of treating them as social partners nor the adoption of the Intentional Stance. As discussed above, it might be either because acquired prior knowledge did not have sufficient time to be consolidated or because it is the first-hand experience of repetitive interaction that makes a difference. The first speculative explanation related to the duration of the formal training is addressed in Section 3.3.2.2, while the interactive aspect is addressed in Chapter 5 on 'Interaction'.

However, Ghiglino et al. (2020) asked 79 participants to observe human-like eye/head movement (copied from actual human data and retargeted on the robot) with a machine-like behaviour where the eyes moved between two locations with the exact same timing and trajectory, thus exhibiting no variability (cf. Fig. 3.2).

One group of participants was told only to observe the robot and to answer half of the IST (see Marchesi et al., 2019) after each observation. Then, this group of participants was asked explicitly whether they detected any difference between the two behaviours, and if the answer was positive, the experimenter asked them which one was more human-like according to them. A second group of participants was told from the very beginning of the experiment that they would have been exposed to two different behaviours, one of which was generated by retargeting a pre-recorded human behaviour. This latter group was also told explicitly that their main task would have been recognizing the human-like behaviour from a more mechanistic one. The findings demonstrated that participants' ability to detect behavioural variability, specifically biological motion, was influenced by their prior knowledge of the behaviours implemented in the robot. When no prior information was provided regarding the nature of the robot's behaviours, participants tended to overlook the details, and nearly half of the participants in the sample failed to

Fig. 3.2 Experimental setup of Ghiglino and colleagues' study (2020), in which participants were exposed to two different behaviours of the robot (human-like vs. machine-like) under two different instructions. (The picture was taken from the original paper by Ghiglino et al. (2020), with the permission of all the authors)

recognise any difference between human-like and machine-like behaviours. Even among those who did notice the distinctions, many incorrectly attributed human-likeness to both types of behaviours. Moreover, there was no significant difference in the likelihood of adopting the Intentional Stance, which was evaluated mainly with the IST (Marchesi et al., 2019) and further explored with the Godspeed questionnaire (Bartneck et al., 2009). In contrast, participants who received explicit instructions regarding the nature of the two behaviours were able to discern differences between them, leading to variations in their perceptions of anthropomorphism, animacy, and likeability towards the robot. Specifically, when participants were directed to pay attention to hints of human-likeness in the robot's behaviours, they displayed more distinct responses on the Godspeed questionnaire between the conditions. It became evident that their perception of anthropomorphism, animacy, and likeability was primarily shaped by their subjective perception of humanness, rather than the objective (human-like/mechanistic) behaviour. This suggests that humans tend to confirm their own biases and anthropomorphize robots based on their perception of human-like behaviour rather than the actual human-like behaviour of the robot. This might be due to the fact that people hold diverse priors regarding the concept of human-likeness, especially when they are asked to interact with a non-human agent (and, possibly, even more if the agent is an artefact).

Overall, the collection of studies presented in this paragraph shows that, on the one hand, prior knowledge about the robot's functionality (delivered in short training sessions) does not influence adoption of the Intentional Stance. However, prior knowledge that allows focusing attention on specific behavioural characteristics of the robot might have an impact on anthropomorphism and the likelihood of adopting the Intentional Stance, but in this case, what matters most is our own preconceptions regarding what behaviours actually constitute 'human' behaviours.

3.3.2.2 Expertise

In the context of the results of short formal training providing unconsolidated knowledge about the robot's inner workings, the obvious question to be asked is: What if participants' prior knowledge about robots has enough time to be consolidated and thus becomes *expertise*, for example, through a long-term education?

When examining the role of education in the adoption of the Intentional Stance, there are two aspects to be considered: the duration and the type of education. Regarding the duration, we might see education as a gradient: individuals might be more or less educated dependent on years of formal training. Regarding the type of education, we might imagine two individuals having the same years of education, but substantially varying in terms of subjects: for example, a background in humanities versus in STEM (Science, Technology, Engineering, and Mathematics). Both duration and type of education are relevant aspects when investigating individuals' attitudes towards robots, which shape further interactions that people would have with them.

Concerning the first aspect (duration of education), Ghiglino and Wykowska (2020) found a negative correlation between the two, namely the shorter participants' duration of education, the more likely they were to adopt the Intentional Stance. As a potential explanation, the authors proposed that individuals with less years of education might have been less exposed to technological knowledge. It would increase the likelihood of adopting the Intentional Stance because naturally, when humans are exposed to unknown (or not easily understandable) systems, such as robots, they would tend to adopt an 'intentional strategy'—the Intentional Stance—as the most available and default model of reasoning about others' behaviour (Marchesi et al., 2021b; Spatola et al., 2022b), as this is the way of reasoning we are most familiar with. Since early childhood, we have been trained in explaining and predicting the behaviours of others with reference to their mental states. Thus, this is the most accessible strategy to explain behaviours of others, and we are likely to use it as a default when explaining behaviours of any system that we do not know how to explain otherwise. Thus, when individuals interpret and explain behaviours of a new and unknown agent, they would be more prone to adopt the Intentional Stance because they are presumably more familiar with the intentional rather than with the mechanical explanations (Spatola et al., 2022b).

However, it is plausible that not only the duration but also the type of education plays a role in the individual tendency to adopt the Intentional Stance towards robots. For example, people with an engineering background are quite likely to adopt the mechanical/Design Stance rather than the Intentional Stance towards robots, as their formal training and perhaps experience in building artificial systems has had long enough duration to consolidate knowledge about the mechanical workings of the robots. On the other hand, individuals with formal training in other (non-engineering) disciplines, since they are less familiar with mechanical (and robotic) technology, might be more guided by the common knowledge they gained through interactions with other humans.

To examine the impact of education type on the likelihood of adopting the Intentional Stance, in their study Roselli et al. (2023) involved a group of individuals with a background in robotics, and another group with a background in psychotherapy. These two groups were chosen to control for the duration of education (comparable across the two groups) while varying the type of education. While obviously roboticists would have an extensive knowledge regarding robot design and programming, psychotherapists would have no professional knowledge about robots, but a long-term training promoting mentalizing skills. In fact, psychotherapists are trained to reason about the mental states of others. Thus, they should have the equivalent expertise regarding the Intentional Stance as roboticists have regarding the Design Stance. Both groups of participants completed the IST (Marchesi et al., 2019), and their neural activity was recorded via EEG during a resting-state phase prior to the task. At a behavioural level, we observed that psychotherapists displayed a greater tendency to ascribe intentionality to robots (i.e. significantly higher IST scores) compared to roboticists. This was in line with the initial hypothesis as long-term training in mentalizing would 'spill over' to the tendency of adopting the Intentional Stance towards others, even if they are non-human agents. On the other hand, having highly specialized training in robotics, and knowing exactly how robots are built and designed, should evoke the adoption of the Design Stance. Notably, participants' behavioural responses were mirrored by their neural activity during the resting state, as demonstrated by higher power in gamma frequency bands (traditionally associated with mentalizing and adoption of the Intentional Stance, Cohen et al., 2009) for psychotherapists compared to roboticists (Roselli et al., 2023).

These results are in line with the findings from another study by Parenti et al. (2023), who examined whether individuals' propensity to adopt the Intentional Stance towards iCub could be related to their patterns of synchrony between various brain regions. To this aim, the authors focused on frontotemporal theta synchronization as it has been shown to be a marker of cognitive flexibility (Fries, 2005). The focus was on low frequencies such as theta, also because they are thought to underlie long-range communication between brain regions during various cognitive processes (Canolty & Knight, 2010; Formica et al., 2022; Solomon et al., 2017), including attribution of mental states and intentions (Bögels et al., 2015; Seymour et al., 2018). For the analyses reported by Parenti et al. (2023), the authors used data collected in Bossi and colleagues' study (2020) as well as data from the experiment of Roselli et al. (2023). This way, the authors compared EEG data from three samples (general population: Bossi et al. (2020); roboticists and psychotherapists: Roselli et al., 2023). The results showed that across all samples, IST scores were positively correlated with synchrony between frontal and left temporal areas in the theta frequency range. In other words, the higher IST scores (i.e. the more likely participants were to perceive iCub as intentional), the higher frontotemporal theta synchrony they displayed (Parenti et al., 2023). This pattern of results is in line with the idea that higher theta synchrony between those regions is linked to mentalizing abilities (Bögels et al., 2015; Seymour et al., 2018).

Overall, the findings described in this paragraph suggest that, in contrast to short-term training, formal educational background and expertise modulate the likelihood of the adoption of the Intentional Stance. This is presumably due to the effect that short-term training can have on knowledge consolidation. Exposure to a 15–30-minute explanation about the robot's mechanical workings does not consolidate deeply enough to modify the representation that individuals have regarding what is a robot. This is obviously different from formal long-term education.

3.4 Conclusions

Individual differences are an important factor that impacts the likelihood of adopting the Intentional Stance, and anthropomorphism more generally. The need for cognition and need for closure influence the propensity to anthropomorphise non-human agents. Furthermore, education (both in terms of duration and type), as well as longer-term interaction with a robot, plays a role in attributing intentional agency to robots. Finally, even if two individuals' traits might not differ substantially in the degree towards which they adopt the Intentional Stance, their state of mind, at a given moment in time, or preconceptions that they have about what constitutes 'humanness' might determine how they approach a robot—with either an intentional or mechanistic mindset.

Interview with Prof. Bertram Malle

by Lorenzo Parenti and Ziggy O'Reilly

When we consider a spectrum of robots, from a basic robot arm to a highly communicative humanoid or android, our responses vary across mental state inferences, emotional reactions, and moral judgments about the robot's actions.

—Prof. Bertram Malle

Prof. Bertram Malle holds a Ph.D. from Stanford University and is currently Professor at Brown University's Department of Cognitive, Linguistic, and Psychological Sciences. Renowned for his research in social cognition, moral psychology, and Human–Robot Interaction, his research has received funding from leading government and academic institutions. With over 170 articles and numerous books to his credit, his work continues to push the boundaries of understanding in the fields of psychology and Human–Robot Interaction.

Summary
On the 5th of January, 2024, two Ph.D. students, Lorenzo Parenti and Ziggy O'Reilly, connected online from Italy (Genova) and Australia (Perth) to speak with Prof. Malle in the United States. The interview covers topics related to the Intentional Stance and Prof. Malle's work on Human–Robot Interaction, trust, and moral agency.

Keywords
Intentional Stance, Moral Stance, Moral Agency

Conversation on the 'Human' Side of the Human–Robot Interaction

Lorenzo Parenti I understand that your research spans various topics, ranging from social cognition to Human-Robot Interaction. Could you share your insights on the interplay between intentionality, or more broadly, the attribution of mental states, and trust, specifically in the context of Human-Robot Interaction?

Bertram Malle Absolutely! The Intentional Stance involves applying the concepts of mental states, such as desires and beliefs, to an agent. Psychologically, it encompasses interpreting behaviour, inferring mental states, and dealing with the uncertainty in explaining behaviour through mental states.

This package varies, and depending on the element we focus on, the relationship to trust may vary in strength.

We conceptualise trust as a state one person has toward the potential future actions of another.

> ***Trust involves expectations of benefitting from the other person's actions, but there's also a concern about vulnerability if they act differently. Trust encompasses various beliefs about the other person, leading to different forms of trust.***

For instance, *reliability, predictability, and consistency* create a basic form of trust—expecting someone to act predictably, like a robot arm performing the same task repeatedly.

On the other hand, there's *competence*, involving not reliability but beliefs in the agent's capacity to handle new situations.

Yet, interpersonal relations introduce a moral dimension to trust. In contexts where actions have a moral value, people use trust to convey expectations about morally correct actions.

This involves mental state inferences related to *integrity*, *sincerity*, and *benevolence*.

Sincerity, in communication, involves trusting that what someone says they believe or will do, they actually believe or will do, creating vulnerability if they don't.

Benevolence, adds another layer, requiring specific intentions related to the other person's well-being.

These facets of trust involve different degrees of mental state inferences, from basic expectations of reliability to more nuanced judgments about moral intentions.

> ***Trust and Intentional Stance are indeed related, but the complexity arises when we delve into specific facets of trust, each requiring understanding of Intentional Stance and mental state inferences.***

Lorenzo Parenti Thank you for your insightful response. I was initially planning to ask about the Multi-Dimensional Measure of Trust (MDMT) questionnaire, but it seems your previous answer covered much of that interesting topic. The questionnaire you developed aims to measure trust towards an agent based on different dimensions. Would it be accurate to say that as the complexity increases, more dimensions come into play, almost like additional layers building upon the previous ones?

Bertram Malle Absolutely, that's a great way to frame it. When we consider a spectrum of robots, from a basic robot arm to a highly communicative humanoid or android, our responses vary across mental state inferences, emotional reactions, and moral judgments about the robot's actions. This encompasses the judgments I previously described—such as inferences about the robot's benevolence and sincerity.

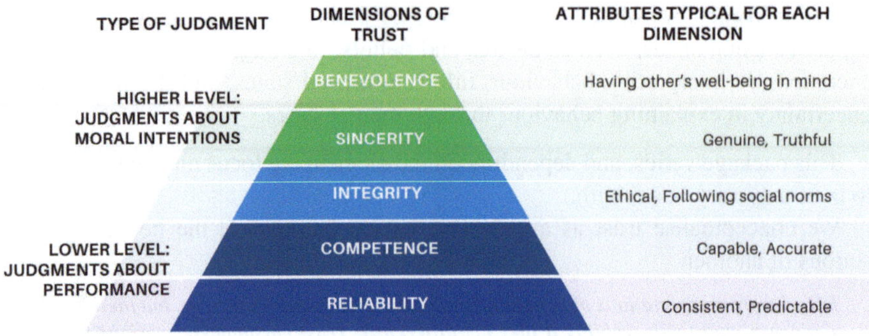

Fig. 3.3 Visual representation of the dimensions of trust and the corresponding trustor's judgements about the trustee's behaviour and mind (including attributions of intentions)

These aspects build on the foundational concerns of reliability, and capability (Fig. 3.3).

The MDMT questionnaire allows individuals to make judgments about the foundational performance aspects of reliability and competence without imposing judgments about moral dimensions, if they are not meaningful.

For instance, when it comes to a robot arm or my car, I do not typically worry about their integrity, sincerity or benevolence. The MDMT recognizes this and provides the option for respondents to label certain questions as "not applicable" or "does not fit." This flexibility enhances measurement accuracy, especially because we do not want individuals to make judgments they would not naturally consider.

This approach becomes an indirect measure of the Intentional Stance. The number of "not applicable" checkboxes reflects how willing one is to adopt a Moral or Intentional Stance toward the other agent.

Interestingly, research indicates that this number tends to increase when dealing with less humanoid or less mentally capable robots, aligning with the Design Stance and Intentional Stance converging.

The less a robot's design appears geared towards interpersonal and social interaction, the more likely individuals are to dismiss certain moral questions, emphasizing the interplay between Design and Intentional Stance.

Lorenzo Parenti Thank you for shedding light on the potential use of the MDMT questionnaire. Moving on to another topic you have explored, I'd like to delve into moral agency in robots and how the attribution of morality influences users' perceptions. From the observer's standpoint, is there a connection between intentionality and morality?

Bertram Malle Certainly, there are two key elements that can intersect.

First, it's crucial to assess whether the robot is designed to make decisions. This involves considering its knowledge, beliefs, and perceptions, along with understanding its goals in specific contexts. Even a robot arm, with its complexity in recognizing, deciding, or repairing, invites a degree of Intentional Stance.

However, for a robot to be considered a moral agent or morally competent, a *second* component is required. Users need to believe that the robot was designed with a set of norms or values guiding its decisions within the moral framework of its community. This goes beyond mere decision-making capacity; it involves making decisions in alignment with moral standards, such as recognizing what's polite, avoiding harm, or respecting another's emotional well-being.

Treating a machine or robot as a moral agent involves both the Intentional Stance and a set of ascriptions related to moral competence and norms.

Moreover, attributing morality to a robot requires not just its understanding of norms but also an ability to express them. This includes using moral vocabulary and justifications, where decisions are explained in terms of prioritizing one norm over another. Such explanations are essential in building a sophisticated and demanding interaction, akin to human relationships where justifications for one's actions are important.

This level of interaction is not yet common with real robots, but through studies, we simulate these scenarios to observe and understand how people respond to complex interactions involving the Intentional Stance and moral attributions.

Ziggy O'Reilly I'd like to clarify the connection between adopting the Intentional Stance and attributing blame to a robot. In your previous explanation, you mentioned the importance of believing that a robot is designed with a set of values and possesses knowledge. Is the belief in the robot having knowledge a necessary component for attributing blame to the robot?

Bertram Malle Yes, I would assert that, especially concerning blame, belief in the robot's knowledge is crucial. While some of our moral judgments may not necessitate the Intentional Stance—for instance, labeling a behavior as morally bad—blame is a different matter.

Consider a scenario where a robot arm unintentionally injures a human worker. While this may be deemed "morally bad" due to the harm caused, blame involves judging more. Blame entails assessing whether the agent intended the action or not, and even if it was unintentional, whether they could have made a better decision.

The very concept of blame presupposes intentional capacities and decision-making abilities that integrate norms into the decision process.

In the case of a machine without intentional capacities and the ability to make moral decisions guided by norms, the term "blame" may not be applicable. In simulated scenarios where people are asked to rate the blame a robot deserves, around 30% of respondents spontaneously express the view that machines cannot be blamed. They might deny the Intentional Stance or the moral competence of the robot, or both. Instead, they might label the behavior as bad or wrong, or express a desire to address the situation through rejection or mistrust, such as reporting issues to the manufacturer.

In summary, the attribution of blame to a robot typically requires both the Intentional Stance and an acknowledgment of the robot's moral competence, including its knowledge of moral norms. Without these components, blame may not be a meaningful or applicable term in the context of human-robot interactions.

Ziggy O'Reilly I appreciate your insights. I recall one of your studies critiquing methodological designs that force participants to choose between dichotomous options (i.e., 'intentional,' and 'unintentional,' actions) when they describe a human action. Would this same critique apply with robots and to experimental design in which participants have to describe robots' actions?

Bertram Malle Absolutely, and perhaps even more pronounced. The critique stands because participants, much like in the human context, may feel coerced to select between intentional and unintentional even if these options do not adequately capture the nuances of their perceptions (e.g., they may want to say the agent acted "willingly" or "knowingly," which mean different things from "intentionally").

However, with robots, there's an additional layer.

First, participants might reject the notion that these two words comprehensively cover the range of descriptions they would use for the robot's actions. They might comply with the forced choice but harbor a sense that neither term perfectly fits.

Second, and more significant, participants might contend that these two words are meaningless when applied to the robot. Unlike humans who can make intentional decisions, robots may lack the capacity for genuine choice and decision-making. For instance, a robot arm executing routine tasks without making choices or decisions may render the distinction between intentional and unintentional actions irrelevant.

People's rejection of the researcher's question becomes even more robust, because they would want to express not only that the options are insufficient but that they are fundamentally inappropriate for the robot.

In designing studies, providing participants with a broader range of descriptive options—such as intentional, not intentional, knowingly, willingly—allows for a more nuanced understanding of their perceptions.

Ziggy O'Reilly Shifting gears a bit, is there a particular social robot that you prefer to work with or contemplate?

Bertram Malle To be honest, such a robot doesn't exist for me.

I do have robots that annoy me more than others, and there are those that raise more concerns, especially when they are misleadingly advertised to possess capabilities they lack.

Currently, my favorite robots are the ones I create within simulated narratives.

In these scenarios, I aim to convince participants that the robot possesses a certain competence, making it plausible to pass moral judgments, provide reasoned explanations for its actions, and distinguish between intentional and unintentional behavior—at least in principle.

When dealing with real present-day robots, complexities arise.

A robot's physical appearance, movements, sounds, and tactile elements introduce challenges for researchers attempting to convey or negate specific capacities to participants. In the lab, handling a physical robot and ensuring it aligns with participants' interpretations is undoubtedly challenging.

Yet, both approaches—the simulated interaction with fictitious robots and the physical interactions with a real robot—represent just the beginning. There is a need to bridge these approaches, with the hope that future advancements allow for the construction of robots with varied appearances, sounds, and tactile qualities to generalise findings. It's a demanding research domain, but the hope is that diverse approaches can yield complementary and convergent conclusions.

If I had substantial resources and someone asked me which robot to acquire for my lab, I wouldn't have a definitive answer. Perhaps building a mixed-reality robot that allows for variations in appearance, sound, gender, and other factors would be more beneficial.

Lorenzo Parenti Certainly, it's a collective effort, and despite the challenges, the field remains fascinating and full of potential. We really appreciate your positive and motivating perspective in this field of research. We also want to thank you, because you gave us some insights regarding studies we are running now related to trust and morality in HRI.

References

Adolphs, R. (1999). Social cognition and the human brain. *Trends in Cognitive Sciences, 3*(12), 469–479. https://doi.org/10.1016/S1364-6613(99)01399-6

Allport, G. W., Clark, K., & Pettigrew, T. (1954). *The nature of prejudice*. Addison-Wesley.

Amodio, D. M., & Frith, C. D. (2006). Meeting of minds: The medial frontal cortex and social cognition. *Nature Review Neuroscience, 7*, 268–277. https://doi.org/10.1038/nrn1884

Bargh, J. A., Gollwitzer, P. M., Lee-Chai, A., Barndollar, K., & Trötschel, R. (2001). The automated will: Nonconscious activation and pursuit of behavioral goals. *Journal of Personality and Social Psychology, 81*(6), 1014–1027. https://doi.org/10.1037/0022-3514.81.6.1014

Barsalou, L. W. (1983). Ad hoc categories. *Memory & Cognition, 11*, 211–227. https://doi.org/10.3758/BF03196968

Bartneck, C., Kulić, D., Croft, E., & Zoghbi, S. (2009). Measurement instruments for the anthropomorphism, animacy, likeability, perceived intelligence, and perceived safety of robots. *International Journal of Social Robotics, 1*, 71–81. https://doi.org/10.1007/s12369-008-0001-3

Bögels, S., Barr, D. J., Garrod, S., & Kessler, K. (2015). Conversational interaction in the scanner: Mentalizing during language processing as revealed by MEG. *Cerebral Cortex, 25*(9), 3219–3234. https://doi.org/10.1093/cercor/bhu116

Bossi, F., Willemse, C., Cavazza, J., Marchesi, S., Murino, V., & Wykowska, A. (2020). The human brain reveals resting state activity patterns that are predictive of biases in attitudes toward robots. *Science robotics, 5*(46), eabb6652. https://doi.org/10.1126/scirobotics.abb6652

Buckner, R. L., Andrews-Hanna, J. R., & Schacter, D. L. (2008). The brain's default network: Anatomy, function, and relevance to disease. *Annals of the New York Academy of Sciences, 1124*(1), 1–38. https://doi.org/10.1196/annals.1440.011

Cacioppo, J. T., Petty, R. E., & Feng Kao, C. (1984). The efficient assessment of need for cognition. *Journal of Personality Assessment, 48*(3), 306–307. https://doi.org/10.1207/s15327752jpa4803_13

Cacioppo, J. T., Petty, R. E., Feinstein, J. A., & Jarvis, W. B. G. (1996). Dispositional differences in cognitive motivation: The life and times of individuals varying in need for cognition. *Psychological Bulletin, 119*(2), 197. https://doi.org/10.1037/0033-2909.119.2.197

Canolty, R. T., & Knight, R. T. (2010). The functional role of cross-frequency coupling. *Trends in Cognitive Sciences, 14*(11), 506–515. https://doi.org/10.1016/j.tics.2010.09.001

Ciardo, F., Ghiglino, D., Roselli, C., & Wykowska, A. (2020). *The effect of individual differences and repetitive interactions on explicit and implicit attitudes towards robots*. In 12th international conference on social robotics. Golden, Colorado. Lecture Notes in Computer Science (Vol. 12483, pp. 466–477). https://doi.org/10.1007/978-3-030-62056-1_39.

Cohen, M. X., David, N., Vogeley, K., & Elger, C. E. (2009). Gamma-band activity in the human superior temporal sulcus during mentalizing from nonverbal social cues. *Psychophysiology, 46*(1), 43–51. https://doi.org/10.1111/j.1469-8986.2008.00724.x

Deligianis, C., Stanton, C. J., McGarty, C., & Stevens, C. J. (2017). The impact of intergroup bias on trust and approach behaviour towards a humanoid robot. *Journal of Human-Robot Interaction, 6*(3), 4–20. https://doi.org/10.5898/JHRI.6.3.Deligianis

Emberson, L. L. (2016). Gaining knowledge mediates changes in perception (without differences in attention): A case for perceptual learning. *The Behavioral and Brain Sciences, 39*, e240. https://doi.org/10.1017/S0140525X15002496

Epley, N., Waytz, A., & Cacioppo, J. T. (2007). On seeing human: A three-factor theory of anthropomorphism. *Psychological Review, 114*(4), 864. https://doi.org/10.1037/0033-295X.114.4.864

Epley, N., Akalis, S., Waytz, A., & Cacioppo, J. T. (2008). Creating social connection through inferential reproduction: Loneliness and perceived agency in gadgets, gods, and greyhounds. *Psychological Science, 19*(2), 114–120. https://doi.org/10.1111/j.1467-9280.2008.02056.x

Fiore, S. M., Wiltshire, T. J., Lobato, E. J., Jentsch, F. G., Huang, W. H., & Axelrod, B. (2013). Toward understanding social cues and signals in human–robot interaction: Effects of robot gaze and proxemic behavior. *Frontiers in Psychology, 4*, 859. https://doi.org/10.3389/fpsyg.2013.00859

Fischer, K. (2011, March). *Interpersonal variation in understanding robots as social actors*. In Proceedings of the 6th international conference on Human-robot interaction (pp. 53–60). https://doi.org/10.1145/1957656.1957672.

Formica, S., González-García, C., Senoussi, M., Marinazzo, D., & Brass, M. (2022). Theta-phase connectivity between medial prefrontal and posterior areas underlies novel instructions implementation. *Eneuro, 9*(4). https://doi.org/10.1523/ENEURO.0225-22.2022

Fries, P. (2005). A mechanism for cognitive dynamics: Neuronal communication through neuronal coherence. *Trends in Cognitive Sciences, 9*(10), 474–480. https://doi.org/10.1016/j.tics.2005.08.011

Fussell, S. R., & Krauss, R. M. (1992). Coordination of knowledge in communication: Effects of speakers' assumptions about what others know. *Journal of Personality and Social Psychology, 62*(3), 378. https://doi.org/10.1037/0022-3514.62.3.378

Garfinkel, S. N., Seth, A. K., Barrett, A. B., Suzuki, K., & Critchley, H. D. (2015). Knowing your own heart: distinguishing interoceptive accuracy from interoceptive awareness. *Biological Psychology, 104*, 65–74. https://doi.org/10.1016/j.biopsycho.2014.11.004

Gazzaniga, M. S. (2004). *The cognitive neurosciences*. MIT Press.

Ghiglino, D., & Wykowska, A. (2020). When robots (pretend to) think. In *Artificial intelligence* (pp. 49–74). Brill Mentis.

Ghiglino, D., De Tommaso, D., Willemse, C., Marchesi, S., & Wykowska, A. (2020). Can I get your (robot) attention? Human sensitivity to subtle hints of human-likeness in a humanoid robot's behavior. *Cogsci* 2020. https://cognitivesciencesociety.org/cogsci20/papers/0168/0168.pdf.

Gollwitzer, P. M., & Sheeran, P. (2006). Implementation intentions and goal achievement: A meta-analysis of effects and processes. *Advances in Experimental Social Psychology, 38*, 69–119. https://doi.org/10.1016/S0065-2601(06)38002-1

Häring, M., Kuchenbrandt, D., & André, E. (2014, March). *Would you like to play with me? How robots' group membership and task features influence human-robot interaction*. In Proceedings of the 2014 ACM/IEEE international conference on Human-robot interaction (pp. 9–16). https://doi.org/10.1145/2559636.2559673.

Henrich, J., Heine, S. J., & Norenzayan, A. (2010). The weirdest people in the world? *Behavioral and Brain Sciences, 33*(2–3), 61–83. https://doi.org/10.1017/S0140525X0999152X

Henschel, A., Hortensius, R., & Cross, E. S. (2020). Social cognition in the age of human–robot interaction. *Trends in Neurosciences, 43*(6), 373–384. https://doi.org/10.1016/j.tins.2020.03.013

Hudson, M., Nicholson, T., Ellis, R., & Bach, P. (2016). I see what you say: Prior knowledge of other's goals automatically biases the perception of their actions. *Cognition, 146*, 245–250. https://doi.org/10.1016/j.cognition.2015.09.021

Isaacs, E. A., & Clark, H. H. (1987). References in conversation between experts and novices. *Journal of Experimental Psychology: General, 116*(1), 26. https://doi.org/10.1037/0096-3445.116.1.26

Jann, K., Kottlow, M., Dierks, T., Boesch, C., & Koenig, T. (2010). Topographic electrophysiological signatures of fMRI resting state networks. *PLoS One, 5*(9), e12945. https://doi.org/10.1371/journal.pone.0012945

Kiesler, S. (2005, August). *Fostering common ground in human-robot interaction*. In ROMAN 2005. IEEE International workshop on robot and human interactive communication, 2005. (pp. 729–734). IEEE. https://doi.org/10.1109/ROMAN.2005.1513866.

Kruglanski, A. W. (1990). Motivations for judging and knowing: Implications for causal attribution. In E. T. Higgins & R. M. Sorrentino (Eds.), *Handbook of motivation and cognition: Foundations of social behavior* (Vol. 2, pp. 333–368). Guilford Press.

Laurent, S. M., Nuñez, N. L., & Schweitzer, K. A. (2015). The influence of desire and knowledge on perception of each other and related mental states, and different mechanisms for blame. *Journal of Experimental Social Psychology, 60*, 27–38. https://doi.org/10.1016/j.jesp.2015.04.009

Lee, S. L., Lau, I. Y. M., Kiesler, S., & Chiu, C. Y. (2005, April). *Human mental models of humanoid robots*. In Proceedings of the 2005 IEEE international conference on robotics and automation (pp. 2767–2772). IEEE. https://doi.org/10.1109/ROBOT.2005.1570532.

Maner, J. K., Kenrick, D. T., Becker, D. V., Robertson, T. E., Hofer, B., Neuberg, S. L., et al. (2005). Functional projection: How fundamental social motives can bias interpersonal perception. *Journal of Personality and Social Psychology, 88*(1), 63. https://doi.org/10.1037/0022-3514.88.1.63

Mantini, D., Perrucci, M. G., Del Gratta, C., Romani, G. L., & Corbetta, M. (2007). Electrophysiological signatures of resting state networks in the human brain. *Proceedings of the National Academy of Sciences, 104*(32), 13170–13175. https://doi.org/10.1073/pnas.0700668104

Marchesi, S., Ghiglino, D., Ciardo, F., Perez-Osorio, J., Baykara, E., & Wykowska, A. (2019). Do we adopt the Intentional Stance toward humanoid robots? *Frontiers in Psychology, 10*, 450. https://doi.org/10.3389/fpsyg.2019.00450

Marchesi, S., Bossi, F., Ghiglino, D., De Tommaso, D., & Wykowska, A. (2021a). I am looking for your mind: Pupil dilation predicts individual differences in sensitivity to hints of human-likeness in robot behavior. *Frontiers in Robotics and AI, 8*, 653537. https://doi.org/10.3389/frobt.2021.653537

Marchesi, S., Spatola, N., Perez-Osorio, J., & Wykowska, A. (2021b). Human vs Humanoid. *A behavioral investigation of the individual tendency to adopt the intentional stance*. In Proceedings of the 2021 ACM/IEEE international conference on human-robot interaction (pp. 332–340). https://doi.org/10.1145/3434073.3444663.

Meyer, M. L. (2019). Social by default: Characterizing the social functions of the resting brain. *Current Directions in Psychological Science, 28*(4), 380–386. https://doi.org/10.1177/0963721419857759

O'Reilly, Z., Roselli, C., & Wykowska, A. (under review). *Does exposure to technological knowledge modulate the adoption of the Intentional Stance towards humanoid robots in children?*. Preprint available at https://psyarxiv.com/cfg4k

Parenti, L., Navare, U. P., Marchesi, S., Roselli, C., & Wykowska, A. (2023). Theta Synchronization as a Neural Marker of flexible (re-)use of socio-cognitive mechanisms for a new category of (artificial) interaction partners. *Cortex, 169*, 249–258. https://doi.org/10.1016/j.cortex.2023.09.018

Powers, A., & Kiesler, S. (2006, March). *The advisor robot: tracing people's mental model from a robot's physical attributes*. In Proceedings of the 1st ACM SIGCHI/SIGART conference on human-robot interaction (pp. 218–225). https://doi.org/10.1145/1121241.1121280.

Roselli, C., Navare, U. P., Ciardo, F., & Wykowska, A. (2023). Type of education affects individuals' adoption of Intentional Stance towards robots: an EEG study. *International Journal of Social Robotics, 16*, 185–196. https://doi.org/10.1007/s12369-023-01073-2

Roselli, C., Marchesi, S., Russi, N. S. et al. (2024). A study on social inclusion of humanoid robots: A novel embodied adaptation of the cyberball paradigm. *International Journal of Social Robotics, 16*, 671–686. https://doi.org/10.1007/s12369-024-01130-4

Schilbach, L., Wohlschlaeger, A. M., Kraemer, N. C., Newen, A., Shah, N. J., Fink, G. R., & Vogeley, K. (2006). Being with virtual others: Neural correlates of social interaction. *Neuropsychologia, 44*, 718–730. https://doi.org/10.1016/j.neuropsychologia.2005.07.017

Seymour, R. A., Wang, H., Rippon, G., & Kessler, K. (2018). Oscillatory networks of high-level mental alignment: A perspective-taking MEG study. *NeuroImage, 177*, 98–107. https://doi.org/10.1016/j.neuroimage.2018.05.016

Solomon, E. A., Kragel, J. E., Sperling, M. R., Sharan, A., Worrell, G., Kucewicz, M., et al. (2017). Widespread theta synchrony and high-frequency desynchronization underlies enhanced cognition. *Nature Communications, 8*(1), 1704. https://doi.org/10.1038/s41467-017-01763-2

Spatola, N., & Chaminade, T. (2022). Cognitive load increases anthropomorphism of humanoid robot. The automatic path of anthropomorphism. *International Journal of Human-Computer Studies, 167*, 102884. https://doi.org/10.1016/j.ijhcs.2022.102884

Spatola, N., & Wykowska, A. (2021). The personality of anthropomorphism: How the need for cognition and the need for closure define attitudes and anthropomorphic attributions toward robots. *Computers in Human Behavior, 122*, 106841. https://doi.org/10.1016/j.chb.2021.106841

Spatola, N., Marchesi, S., & Wykowska, A. (2022a). Different models of anthropomorphism across cultures and ontological limits in current frameworks the integrative framework of anthropomorphism. *Frontiers in Robotics and AI*, 230. https://doi.org/10.3389/frobt.2022.863319

Spatola, N., Marchesi, S., & Wykowska, A. (2022b). Cognitive load affects early processes involved in mentalizing robot behaviour. *Scientific Reports, 12*(1), 14924. https://doi.org/10.1038/s41598-022-19213-5

Spatola, N., Marchesi, S., & Wykowska, A. (2023). The phenotypes of anthropomorphism and the link to personality traits. *International Journal of Social Robotics, 15*, 3–14. https://doi.org/10.1007/s12369-022-00939-1

Spunt, R. P., Meyer, M. L., & Lieberman, M. D. (2015). The default mode of human brain function primes the intentional stance. *Journal of Cognitive Neuroscience, 27*(6), 1116–1124. https://doi.org/10.1162/jocn_a_00785

Sundar, S. S., Bellur, S., Oh, J., Jia, H., & Kim, H. S. (2016). Theoretical importance of contingency in human-computer interaction: Effects of message interactivity on user engagement. *Communication Research, 43*(5), 595–625. https://doi.org/10.1177/0093650214534962

Tanyi, R. A. (2002). Towards clarification of the meaning of spirituality. *Journal of Advanced Nursing, 39*(5), 500–509. https://doi.org/10.1046/j.1365-2648.2002.02315.x

Thellman, S., & Ziemke, T. (2019). *The Intentional Stance toward robots: Conceptual and methodological considerations*. In The 41st Annual conference of the cognitive science society, July 24–26, Montreal, Canada (pp. 1097–1103). Cognitive Science Society.

Urquiza-Haas, E. G., & Kotrschal, K. (2015). The mind behind anthropomorphic thinking: Attribution of mental states to other species. *Animal Behaviour, 109*, 167–176. https://doi.org/10.1016/j.anbehav.2015.08.011

Vogeley, K., & Fink, G. R. (2003). Neural correlates of the first-person-perspective. *Trends in Cognitive Sciences, 7*, 38–42. https://doi.org/10.1016/S1364-6613(02)00003-7

Webster, D. M., & Kruglanski, A. W. (1994). Individual differences in need for cognitive closure. *Journal of Personality and Social Psychology, 67*(6), 1049–1062. https://doi.org/10.1037/0022-3514.67.6.1049

Williams, K. D., & Jarvis, B. (2006). Cyberball: A program for use in research on interpersonal ostracism and acceptance. *Behavior Research Methods, 38*, 174–180. https://doi.org/10.3758/BF03192765

Chapter 4
Factors Influencing the Adoption of the Intentional Stance: The Robot

Davide Ghiglino, Lorenzo Parenti, and Abdulaziz Abubshait

Baymax: On a scale of one to ten, how would you rate your pain?
Hiro: Zero.
Baymax: It is alright to cry.
Hiro: No! No, no, no, no, no!
Baymax: Crying is a natural response to pain.
Hiro: I'm not crying.
Baymax: I will scan you for injuries.
Hiro: DON'T scan me.
Baymax: Scan complete.

– Williams, C., Hall, D. [Directors], 2014, Big Hero 6 [Film], Walt Disney Animation Studios.

4.1 Introduction

It takes two to a tango … Obviously, when discussing the factors contributing to the likelihood of adopting the Intentional Stance, not only the human, but also the robot—the way it looks and the way it behaves—plays a major role. Already during the design phase, engineers commonly endow robots with functions aimed at

D. Ghiglino (✉) · A. Abubshait
Italian Institute of Technology, Social Cognition in Human-Robot Interaction Unit, Genova, Italy
e-mail: davide.ghiglino@iit.it

L. Parenti
Italian Institute of Technology, Social Cognition in Human-Robot Interaction Unit, Genova, Italy

Department of Psychology, Università di Torino, Turin, Italy
e-mail: lorenzo.parenti@iit.it

A. Wykowska (ed.), *Intentional Stance Towards Humanoid Robots*,
https://doi.org/10.1007/978-3-031-65483-1_4

maximizing the naturalness of interaction, which often means human-like characteristics.

Humanoid robots are a particular case of human-like features, both in terms of appearance and motor repertoire. Still, whether an artificial agent that can look and behave like a human is capable of boosting the spontaneity and naturalness of interaction is an open question. In particular, it is still a matter of debate whether artificial agents embedded with 'enough' human-likeness could lead people to treat them as 'thoughtful' or 'intentional' agents. Indeed, we have plenty of examples of artificial agents with different features, but only a few attempts at systemising which of the above-mentioned factors related to the robot design affect humans' tendency to anthropomorphize them (Varella, 2018). In this chapter, we will examine some of the aspects of robot design that might influence the perception of the user. In particular, we will focus on the factors that we claim to be more relevant to adopt the 'Intentional Stance' towards robots, a construct that we defined in the previous chapters as the disposition to treat an entity as an agent possessing mental states, which can be used to interpret and predict the behaviour it displays.

Based on past research, this chapter will focus on the following factors: robot appearance (Sect. 4.1), human-based behaviours (Sect. 4.2), biologically inspired kinematics (Sect. 4.3), and responsiveness and predictability to environmental events (Sect. 4.4).

4.2 Robot Appearance

The physical appearance of a robot is often the first feature that influences human perception (Zlotowski et al., 2016). Various design choices, such as shape, size, and proportions of the robot body, can elicit different responses from humans (Rosenthal-Von Der Pütten & Krämer, 2014). Research has shown, not surprisingly, that robots with a more human-like face, including expressive eyes, eyebrows, and mouth, tend to receive greater human-likeness attributions. In fact, human-like faces are one of the key elements of physical appearance that not only attract our (human) attention but also tend to increase mind attribution (Intentional Stance) and life attribution (animism) (e.g. Looser & Wheatley, 2010). Additionally, anthropomorphic features such as limbs, fingers, and joints contribute to the perception of human-like qualities (Epley et al., 2007).

Indeed, the human brain uses highly sensitive mechanisms that are specialised to detect a broad spectrum of signals (i.e. facial expressions, changes in gaze direction) crucial during social interactions. These social signals are in many cases behavioural manifestations of mental states. For example, when we see a person smile, we infer that he or she is happy. Such signals are hints that increase or decrease the likelihood of adopting the Intentional Stance. Thus, identifying how these specific parameters facilitate adopting the Intentional Stance is crucial. Research in Human–Robot Interaction (HRI) reveals that implementing human-like characteristics in artificial agents facilitates social interaction (Fink, 2012). Robots that show human-like design and behaviour are better accepted (Duffy, 2003), perceived as more pleasant (Axelrod & Hone, 2005) and more usable (Riek et al., 2009), are easier to get acquainted with (Hegel et al., 2008), and are more engaging (Bartneck & Forlizzi, 2004), relative to purely functional designs. Indeed, the physical appearance of a robot can immediately evoke a sense of familiarity and enhance the perception of the robot as a social agent, making it more trustworthy, appreciable, and more engaging, when compared to machines lacking such features (Perez-Osorio & Wykowska, 2019).

However, it is important to pose a question of the extent to which we desire human-likeness in non-human objects, and how people will respond to robots that resemble humans in their physical characteristics. Some authors have argued that a physical appearance of a robot that is too close to that of a human might trigger fear reactions and might be perceived as threatening, due to blurring the distinctiveness of human/robot categories (e.g. Ferrari et al., 2016; Kaplan, 2004; MacDorman et al., 2008). In 1970 the Japanese roboticist Masahiro Mori coined the term 'Uncanny Valley' (UV) describing this interesting relationship between the physical appearance of robots and human perception (see Mori et al., 2012). The idea behind the UV is that the more robots (or artificial beings in general) become human-like in their appearance and behaviour, the higher the increase in the positive emotional response from humans. However, at a certain point, when the artificial being becomes almost, but not quite, indistinguishable from a real human, a sudden drop in emotional response occurs. This drop manifests a sense of unease, eeriness, or even repulsion in humans, hence the term 'uncanny'. The UV can be attributed to several factors. One of the key factors is the human tendency to perceive and interpret subtle cues that indicate the 'typical' human behaviour. When a robot or artificial being approaches human-likeness but falls short of being exactly like a typical human, these discrepancies can create a sense of cognitive dissonance, leading to a negative emotional response. This could be due to factors such as unnatural facial expressions, stiff or jerky movements, or a lack of eye contact. This phenomenon suggests that there is a delicate balance to be struck when designing robots or artificial beings to interact with humans. It might actually be more beneficial to stay within the realm of clearly non-human design.

In the past, researchers have utilized morphing techniques as a key experimental manipulation to explore the phenomenon of the UV. These techniques enable the creation of stimuli that exhibit varying degrees of human-likeness in robot appearance, providing a continuum between extreme robot appearance and fully human appearance. While earlier studies provided limited evidence for the UV, more recent research has shed more light on this phenomenon and its potential underlying causes.

For instance, a study by Abubshait and Wiese (2017) demonstrated that subjective ratings of likeability, mind perception, and trust followed the hypothesized pattern of the UV. Further investigations delving into this phenomenon revealed that the UV effect also influences behaviour, specifically in terms of orienting attention when individuals encounter an 'uncanny' robot. Abubshait and colleagues examined further this finding in a subsequent study (2020), reporting results suggesting that this effect arises due to a depletion of cognitive resources as individuals grapple with the task of categorizing the robot—whether as human or non-human—a phenomenon consistent with the categorical perception hypothesis proposed by Cheetham (2017). In their study, Abubshait and colleagues examined the impact of categorical ambiguity on the ability to top-down control social attention, hypothesizing that faces with categorical ambiguity, specifically those that can be categorized as either 'human' or 'non-human', would generate cognitive conflict. This conflict would potentially deplete cognitive resources. The authors reported three experiments to explore these effects, which confirmed that categorical ambiguity indeed led to cognitive conflict, and that the control over attentional orienting in response to gaze varied based on the level of categorical ambiguity exhibited by the gazing face, and that this variation correlated positively with the processing of cognitive conflict.

Other paradigms have demonstrated similar effects, supporting the interpretation that cognitive conflict arises from categorizing ambiguous faces. For instance, Wiese et al. (2019) conducted a vigilance task measuring the depletion of cognitive resources and found that embedding categorically ambiguous faces in the task resulted in faster resource drainage compared to unambiguous faces. Crucially, they discovered that pre-exposure to ambiguous faces reduced cognitive resource consumption over time in the vigilance task. This finding indicates that pre-exposure diminishes the drain on cognitive resources by categorically ambiguous faces (Wiese et al., 2019). Moreover, longer exposure to novel (ambiguous) stimuli allows for more thorough processing of perceptual features, leading to freeing up of cognitive resources (Desimone & Duncan, 1995; Ferrey et al., 2015). Thus, once the novelty of categorically ambiguous faces diminishes due to increased exposure time, the drainage of cognitive resources resulting from the competition between categorical representations is also reduced. This interpretation is supported by neural data indicating that the activation of fewer neurons in the inferior temporal cortex occurs when stimuli become more familiar to primates, allowing more resources to be allocated for visual processing (Vaitonyte et al., 2023).

Overall, this body of work shows that categorical ambiguity in a robot's appearance seems to blur the distinction between human and non-human traits. Such a conflict triggers the UV effect as the cognitive system struggles to reconcile the ambiguous categorization, which creates a sense of unease and discomfort. Empirical research is still needed to determine whether such cognitive conflict and a sense of unease affect the adoption of the Intentional Stance, and if so, what would be the direction of the effect.

In sum, human-like appearance in robots, with human-like features (face specifically), certainly are shortcuts to anthropomorphism and increase the likelihood of life- or mind attribution. However, future research still needs to address the advantages (positive rapport, smoothness of interaction) and disadvantages (fear, cognitive conflict, and dissonance) of appearance that is perhaps too human-like and might blur the category boundary between who is a human and what is a machine. The latter point is particularly important for ethical guidelines and legal frameworks regarding future robot design.

4.3 Human-Based Robot Behaviours

Even though physical (human-like) appearance seems to be a key determinant of perceiving a robot as human-like (and thus, presumably, to adopt the Intentional Stance towards it), it is certainly not the sole determinant, and perhaps can even be overridden by other factors, such as the robot behaviour. Humans naturally tend to attribute mental states and intentions to non-human agents (even simple geometrical figures such as those presented in video clips of Heider & Simmel, 1944), when the agents display goal-oriented behaviour or human-like behavioural patterns. Some scholars argue that implicit signals exhibited through simple behaviours automatically lead humans to perceive an agent's behaviour as reflecting mental operations (Frith & Frith, 2006). Robots, therefore, should be a particular case, as they are physically embodied and thus have the capacity to act on the environment (potentially in a human-like manner) that might be the most important factor for attribution of intentionality.

One crucial characteristic that defines human-like behaviours is variability (Gielniak et al., 2013). Behavioural observations have shown that humans never exhibit the exact same behaviour twice. For instance, studies have demonstrated that individuals tend to adopt unique patterns of kinematic strategies even when attending to the same target (see, for example, Freedman & Sparks, 2000). This is partially in line with the results of a recent study conducted by Ciardo et al. (2022) that investigated the role of behavioural variability in Human–Robot Interaction. The study involved an interactive task where participants engaged in joint action with the humanoid robot iCub, and the robot's behaviour was manipulated to be either pre-programmed or teleoperated by a human partner. The results showed that humans were sensitive to human-like response variability, as they were able to detect the human-controlled condition above chance, even though the cues they could use for this judgement were quite minimal (the robot was only repetitively pressing a key in response to a stimulus presented on the screen and participants could judge the humanness only based on the distribution of the reaction times of the key presses across the experiment). This finding suggests that human-like variability in even simple behaviours is one of the important parameters for attribution of humanness to other (also artificial) agents and that our brain is sensitive to such subtle characteristics.

Ghiglino et al. (2020a, b) tasked participants with observing a video featuring an avatar (either human or humanoid, cf. Fig. 4.1) embodying various eye movement patterns classified as 'mechanistic' or 'human-like', where the latter simulated the act of reading. Importantly, the avatar eye movements were based on recordings of human eye movements, measured by means of an eye-tracking device—specifically, a Tobii Pro eye tracker (Tobii.com). During the recording sessions, the human participant was presented with textual content displayed on a monitor, prompting them to actively engage in the process of reading. This endeavour was undertaken to measure human-like eye movements (to be later implemented on the avatar). Conversely, the mechanistic condition was construed by employing a distinct

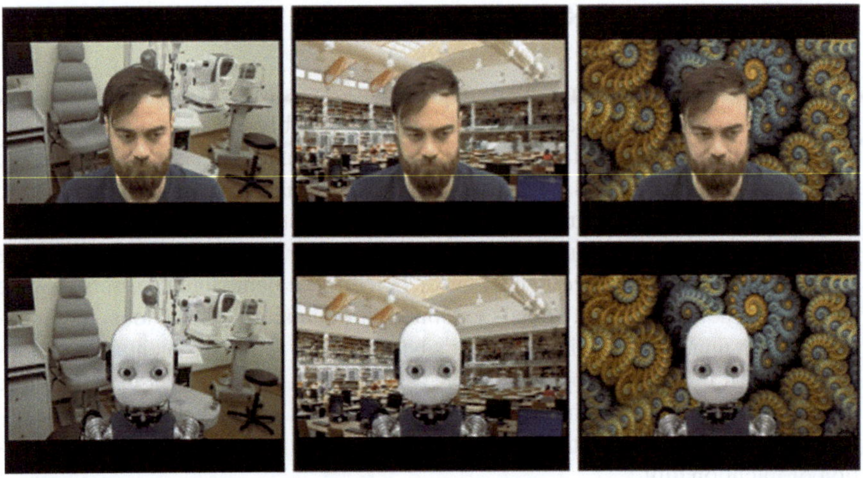

Fig. 4.1 Virtual avatars of the study by Ghiglino et al. (2021) presented in different background contexts

strategy—a dot was introduced onto the screen, following a lateral trajectory between two points, and the participant was asked to pursue this dot with their gaze while their eye movements were recorded, and later implemented on the robot avatar in the 'mechanistic' condition. As the eyes moved repetitively between two discrete points in space, this made the impression of a mechanistic pattern of eye movements.

Once the eye movements of the human participant were implemented on the avatar, Ghiglino and colleagues set out to conduct their actual experiment where participants' primary task was to discern whether the observed eye movements were characteristic of being 'calibrating cameras/mechanistic movements' or 'reading/human-like movements'.

The researchers collected decision-making times, accuracy metrics, and eye-tracking throughout the course of this experiment. While these data did not exactly measure perceptions of a robot's ability to be intentional, the authors measured implicitly attributions of intentionality, in the form of fixation times and the probability of gaze-following, as well as subjective ratings of human-likeness (i.e. a construct adjacent to attribution of intentionality).

The results showed that individuals needed less attentional effort to process and interpret an artificial agent's behaviour when it closely resembled that of a human. Interestingly, the more 'variable' behaviour of reading received always higher ratings on anthropomorphism compared to more 'constant' or 'fixed' mechanistic behaviour.

Another study by Ghiglino and Wykowska (2020) presented participants only with the iCub avatar and the eye movements were only between two points in space. However, the authors manipulated the variability and speed of the eye movements so that the movements could have either fixed tempo or a slightly variable tempo, within human-like, or outside human-like (slow), range of temporal parameters of eye movements. The results showed that slow but variable robot eye movements yielded higher human-likeness ratings, relative to fixed (no variability) eye movements. Interestingly, the eye-tracking data suggest that the human range of eye movement trajectory times were most engaging and evoked spontaneous involvement in joint attention, as measured by the observers' fixation durations on the robot eye region and gaze following. This suggests an interesting dissociation between implicit markers of attunement with human-like artificial agents and explicit reports related to attribution of human-like characteristics. It might be the case that even when we explicitly judge a given behaviour as seemingly human-like, we might still attune better (at an implicit level) to robot behaviours that have indeed human-like characteristics. Most importantly, this study provides yet another piece of evidence that variability even in subtle behaviours such as eye movements is an important hint to the observers regarding humanness.

In another work Ghiglino et al. (2021) attempted to implement eye movements recorded from a human in the physically embodied iCub robot. In contrast to implementing the human saccadic behaviours on avatars, implementing pre-recorded human movements on a physically embodied robot turned out to be quite challenging, as the eye movements needed to be retargeted (adapted to the motor repertoire of the humanoid robot iCub).

It is of note here to point out that retargeting a pre-recorded behaviour might not be the most efficient strategy to generate biologically inspired or human-like behaviour in robots. The first limit of this approach is that it is, by nature, extremely time-consuming. First, one needs to define the behaviour of interest among the entire spectrum of behaviours our species can display. That specific behaviour needs to be recorded in an ecological environment so that the 'model' displays it in a way that is as natural as possible. Among a (possibly) large number of recorded behaviours from different humans, the researcher needs to extract the main features (i.e. temporal vs. spatial information, low vs. high variability, etc.) to consider for maximizing the realistic nature of the behaviour, when retargeted on the robot. Based on an iterative process that takes into account the capabilities of the robot (e.g., the difference between the degrees of freedom of the human who generated the behaviour and the ones of the robot, or the maximal speed that the robot can achieve in that movement) one final behaviour is eventually chosen and adapted to the robot. Once the final human-like behaviour is embedded in the robot, a control condition of a non-human-like behaviour should be defined to validate the approach (e.g. by, for example, asking observers to classify the human-like behaviour as indeed perceived as human-like and the non-human-like as non-human-like) and hopefully demonstrate that a biologically inspired behaviour can improve Human–Robot Interaction, in contrast with the machine-like one.

This process can be feasible with very specific behaviours, given very specific contexts, making it a robust approach from a methodological point of view, that can be adopted in fundamental research to study cognitive processes that are put in place during different kinds of Human–Robot Interaction, and maybe for improving future robots' design (Wiese et al., 2017). However, for more applied purposes, such as defining the best behaviour a commercial robot should display at an airport or at a mall to greet customers, this procedure might be too costly.

A second limit of this approach is that reproducing biologically inspired motion in an artificial agent's body can be challenging due to the structural differences between the human body and a robot body. Here are some reasons why these differences might make this approach extremely challenging (see, for example, Yamane & Murai, 2018; Padois et al., 2017):

1. Skeleton and joint configuration: The human body has a complex skeletal structure with numerous joints that allow for a wide range of motion. Robots, on the other hand, typically have rigid bodies with simple joint mechanisms. Replicating the intricacies of human joints and their movements in a robot body is technically complex and often requires advanced engineering solutions that are difficult to achieve.
2. Muscle systems: Human motion is enabled by the coordinated contraction and relaxation of muscles. Muscles provide flexibility, compliance, and the ability to generate complex forces required for various tasks. Robots, however, typically use motors or actuators that have limited flexibility and force generation capabilities. While there have been advancements in soft robotics and biomimetic materials, replicating the functionality of human muscles in an artificial agent remains a significant challenge.
3. Sensory feedback: The human body is equipped with a sophisticated sensory system that provides continuous feedback about the environment, body position, and forces applied. Proprioception, the ability to sense the position and movement of body parts, is crucial for coordinated motion. Robots often rely on external sensors such as cameras or range finders, which are not as precise or comprehensive as the human sensory system. The lack of accurate and real-time sensory feedback in robots can make it difficult to achieve biologically inspired motion.
4. Energy efficiency: Human bodies are designed to be energy-efficient, utilising muscle flexibility and tendons to store and release energy during movement. This energy recycling mechanism helps minimise energy consumption. Robots, on the other hand, often rely on direct energy input, which can be less efficient and require more power to achieve similar movements. The energy requirements and limitations of robots make it challenging to reproduce the intricate and efficient motion observed in biological systems.
5. Neural control: The human body's motion is governed by a complex network of neurons that control muscle activation and coordination. The human brain and spinal cord provide sophisticated control mechanisms that adapt and learn over time. While artificial intelligence and machine learning techniques have advanced in controlling robot motion, replicating the adaptability, learning and neural control exhibited by humans remains a formidable task.

Given these intrinsic differences, using animation principles to develop biologically inspired and human-like behaviours might be an alternative approach that can overcome some of the challenges associated with directly recording human behaviour.

In fact, by leveraging the principles used in animation production, it is possible to create behaviours that are realistic and evoke a sense of biological (or human-like) motion. These principles can be derived from the observation of human motion and can be used to inform the design and implementation of artificial agent behaviours. Some animation principles that are commonly employed include:

1. Squash and stretch: This principle involves exaggerating the shape and deformation of objects or characters during movement to convey a sense of weight and flexibility.
2. Anticipation: By incorporating anticipatory movements at the beginning of an action, the behaviour appears more natural. This principle helps create a sense of preparation before seeing the movement executed.
3. Timing and spacing: The timing and spacing of different body parts during a motion sequence contribute to the perception of realistic motion. By understanding the natural timing and spacing of human movements, artificial agents can be programmed to replicate these characteristics.
4. Arcs: Most human movements follow curved paths rather than straight lines. Incorporating arc-like trajectories in artificial agent behaviours can enhance their naturalness and biological realism.
5. Follow-through and overlapping action: This principle involves incorporating secondary motions that continue after the primary action has stopped. It adds a sense of naturalness and smoothness to the behaviour.

These are only a few examples of the 12 principles that are defined in the study of 2D and 3D animations (see Lasseter, 1998; Thomas & Johnston, 1981). By leveraging these animation principles, researchers and designers can craft robot behaviours that draw inspiration from biological movement patterns and exhibit a human-like quality, all without needing to directly replicate recorded human actions. Such approaches can greatly enhance the smoothness and authenticity of HRI, making interactions feel more intuitive and relatable to humans. Importantly, the incorporation of these principles is likely to also have significant implications for how intentions are attributed to robots. When robots display behaviours that resonate with the natural motion patterns observed in human actions, human observers are more likely to perceive the robots as having intentional agency. It is plausible to assume that this is due to the mirror neuron system in humans, which simulates observed actions and facilitates the understanding of others' intentions, thereby resonating with actions of others (Gallese et al., 1996). Therefore, by aligning robot behaviours with biologically inspired animation principles, designers can indirectly influence the attribution of intentions by fostering a stronger sense of predictability and familiarity of the robot movements. This alignment between motion and intention perception may not capture the full complexity of human behaviour and motor control, but it offers a practical alternative that can be applied in various domains, including robotics, animation, and virtual characters.

Concerning mirror neuron system (MNS)—a system of specialized neuronal cells that are activated both when an individual performs a particular action and when they observe someone else performing the same action (Gallese et al., 1996;

Di Pellegrino et al., 1992; for a recent review, see Bonini et al., 2022)—it has been postulated to underlie motor resonance that is automatically activated when humans (or, initially, non-human primates as in Pellegrino et al., 1992) observe actions of others. Motor resonance effects have been observed during observation of robot actions if a robot shared a similar motor repertoire to a human (Bisio et al., 2014; Chaminade et al., 2007; Cross et al., 2011; Gazzola et al., 2007; Kilner et al., 2003; Kupferberg et al., 2012). For example, Gazzola et al. (2007) found—in a study using functional magnetic resonance imaging (fMRI)—that brain regions of the human motor resonance network (more specifically, parietal and premotor regions) were activated in response to observation of both human and robot hands grasping an object. Similarly, Cross et al. (2011) showed that observing a simple Lego robot performing dancing movements activated the motor resonance system (bilateral inferior parietal lobules). Most interestingly for the main theme of this book, Chaminade and colleagues (Chaminade et al., 2010) have shown that responses of the motor resonance system were dependent on the task given to participants. In particular, higher responses were observed for observed robot versus human actions in a condition in which participants were asked to judge a mental state (emotion) behind a gesture, relative to when judgements were made based on other features of the movement (e.g. quantity of the movement). This was interpreted as higher involvement of the motor resonance system when processing the robot movements as intentional, compared to movements that are not perceived in intentional terms. This clearly suggests a close link between the motor resonance mechanism and the Intentional Stance. In consequence, as mirror neurons might be an underlying mechanism for action and intention understanding, human-like motor repertoire of robots might indeed facilitate inference of intentions (from observed behaviour), and, as a result, presumably, adoption of the Intentional Stance.

In sum, the perception of human-likeness in a robot's behaviour seems to be influenced by its behavioural characteristics, encompassing factors such as human-like temporal parameters, variability, and movement kinematics (Gielniak et al., 2013). Activation of mirror neurons—through, for example, biologically inspired animation principles—might facilitate the inference of action goals and intentions exhibited by robots. However, there is a delicate balance to strike, as a humanoid robot that perfectly mimics human behaviour can elicit a shift towards the UV phenomenon (Mori, 1970) described in the previous section.

Of note is an interesting dissociation found in some of the studies reported above: a dissociation between implicit markers of engagement (which are influenced by truly human-based behavioural patterns) and explicit judgements of human-likeness which are rather influenced by the mental representation that we have of what 'human-like' behaviour is. Thus, when examining the impact of robot's behaviour on the human observer, one needs to address both the implicit and explicit levels of perception.

4.4 Biologically Inspired Movement Kinematics

As exemplified by popular Disney or other animated movies, the principles of animations can be applied to different types of agents (animals, humans, robots) and objects (lamps, candles). These entities differ in terms of structural characteristics (e.g. shape and size) and kinematics. The structure and kinematics of a moving body are closely intertwined. The dimensions, masses, and physical properties of a body play a crucial role in determining its dynamics and the forces acting upon it, which ultimately influence its kinematics or motion patterns. In the context of designing behaviour for animated characters, it's important to consider both the structural characteristics of the character's body and its kinematics. These aspects contribute to the overall perception and realism of the character's movement.

However, research has shown that it is possible to generate point-light displays of biological motion that selectively emphasize either the structural or kinematic domain. For example, in a study by Mather and Murdoch (1994) involving artificial point-light walkers, the authors found that certain structural features, such as the 'shoulder/hip width ratio', were less informative for determining the sex of a walker compared to kinematic features like 'body sway'.

This suggests that certain features related to the structure of a body may not be as crucial for perceiving or categorizing biological motion as the kinematic aspects of the movement itself. It highlights the importance of understanding the specific features and information that contribute to the perception of biological motion in different contexts.

Kinematics is a powerful characteristic that can evoke perceptions of animacy and intentionality in both animated agents and biological agents. This concept has been supported by seminal works such as those by Heider and Simmel (1944) and Michotte (1946/1963), as well as more recent research.

For example, Cavallo et al. (2016) explored whether movement kinematics can serve as a foundation for intention detection in a simple motor action (such as grasping to pour vs. grasping to drink). Employing a blend of kinematic analyses, psychophysical methods, and classification and regression tree (CART) modelling, the authors revealed that observers utilise a distinct set of discriminant kinematic features within the broader kinematic pattern to discern intentions when observing basic motor actions. The ability to distinguish intentions in their experiments varied from trial to trial, based on movement kinematics, and correlated with the presence of distinctive features in the observed movements. Their results underscore a quantifiable and measurable link between specific features in observed movements and the capacity to differentiate intentions.

In a subsequent study by Koul et al. (2019), the authors explored how prior anticipations of others' intentions merge with visual movement patterns over time to detect the intention behind an observed motor action (such as grasping to pour vs. grasping to drink). Employing rigorous psychophysics techniques, their study revealed that the cognitive process of attributing intentions to others conforms well to drift-diffusion models, where evidence from observed movements accumulates until a decision threshold is met. Comparative assessments of competing models indicate that in situations where movement kinematics don't hold intention information, prior expectations shape the observer's intention attribution. However, when movement kinematics convey intention information, these patterns guide the intention attribution. These findings illustrate a diffusion process in which the impact of expectations is moderated by the informativeness of movement patterns, showing that informative kinematics can override initial expectations.

These studies have direct relevance to HRI, particularly when it comes to humanoid robots that aim to mimic human-like kinematics in movements. It is also important to highlight the interplay between prior expectations and movement kinematics in attributing intentions, which becomes crucial in HRI scenarios, as humans often come into interactions with preconceived expectations of how robots should behave, even at a kinematic level. This goes in line with the discussion presented in Chapter 3 on the human prior biases and attitudes in relation to robots, when it comes to adoption of the Intentional Stance.

Nonetheless, it is important to unfold something that the reader might have already grasped at this point of the book: the more we navigate HRI scenarios the more the boundaries between the human, the robot, and the interaction between the two seem to blur. This implies that even when exploring biologically inspired movement kinematics on the robot's side, one must always take into consideration the perspective given by the human. Indeed, any form of interaction implies the generation of something that goes beyond the mere sum of the parts (or parties) involved, a topic that will be explored in Chapter 5.

In addition to the mere movement kinematics, there are other 'patterns' of behaviour that the human brain detects and interprets in various ways (e.g. movement of a human vs. animal; movement of a man vs. a woman, movement of my friend vs. a stranger). We are referring here to what has been termed 'biological motion' (Johansson, 1973).

In relation to robots, Parenti et al. (see Parenti et al. 2021, 2022), created a 3D virtual avatar of the robot iCub and animated it to exhibit behaviours that incorporated both biological motions based on animation principles and machine-like movements. By utilising a virtual avatar, the authors could also manipulate the appearance of the avatar (a robot vs. a human) while maintaining motion patterns consistent. This enabled the authors to disentangle the effects of appearance from the specific types of motion displayed by the avatar. To generate behaviours that followed biological motion, they employed animation principles inspired by natural movements observed in humans. These principles allowed them to replicate certain kinematic patterns, such as fluidity, coordination, and subtleties in motion, which are typically associated with biological agents. In contrast, the machine-like movements were designed to emulate robotic actions characterized by more rigid and jerky motions. By incorporating both types of movements within the animated behaviours, Parenti et al. aimed to examine the influence of appearance on the perception of biological motion and the potential differences in the way observers responded to the avatar's actions. The findings from these experiments revealed a notable accuracy among participants in categorising behaviours based on their features. Specifically, participants tended to rate behaviours inspired by mechanical motion as more machine-like, while behaviours inspired by biological motion were perceived as more human-like.

In general, from simple geometric shapes, such as a triangle as used in Heider and Simmel's experiments, to point-light stimuli with human-like body structures (see Atkinson et al., 2012), the human-like motion can be discriminated effortlessly from motion of inanimate objects. Several factors come into play, including motor experience, visual experience, and emotional processes, which can differentiate the perception of animated or observer-like entities from inanimate entities (Shiffrar & Thomas, 2013).

Motor experience refers to an individual's personal experience with bodily movements and actions. Humans have a rich motor repertoire and are highly attuned to perceiving and understanding human motion based on own motor experiences, and the activation of the mirror neuron system, as discussed above.

Visual experience also plays a role in the perception of motion as human-like. Through visual exposure to human movements in everyday life, individuals develop perceptual expertise in detecting and interpreting human motion cues. This expertise contributes to the differentiation between human motion and the motion of inanimate objects.

Furthermore, emotional processes play a significant role in shaping how a movement is perceived. Emotions are crucial for movement perception in a twofold manner. First, emotions can be conveyed easily through biological motion by the actor (see Lammers et al., 2019). Second, emotional responses to human-like appearances and movements can enhance the perception of animacy and intentionality. Indeed, human-like motion, when effectively simulated in robots, has the remarkable ability to evoke emotional responses in human observers (Reuten et al., 2018). This phenomenon is rooted in the way humans are wired to interpret and engage with their social environment. As noted in the previous section of this chapter, when we encounter movements that resemble those of living beings, our brain's mirror neuron system activates, allowing us to implicitly understand and simulate those actions. This activation might trigger also an 'empathetic' response, as if we were subconsciously projecting our own experiences onto the observed entity, be it human or robot. This empathetic response is often accompanied by a sense of social resonance—a feeling that we are interacting with something that shares characteristics with living creatures. In the context of Human–Robot Interaction, this emotional connection is pivotal as the perception of human-like motion in robots can foster a sense of familiarity and relatability. When robots move in ways that align with our expectations, based on our observations of human behaviour, it becomes easier for us to attribute intentions to their actions. The more a robot's movement resembles human actions, the more likely we are to intuitively infer the intentions behind its behaviour, much like we do with fellow humans. In essence, the emotional response triggered by human-like motion in robots serves as a bridge between the mechanical and the human-like. It blurs the boundary between the two and can lead us to perceive the robot as having qualities akin to living beings, capable of intentions and agency. As a result, designing robots with animations that evoke these emotional responses becomes a powerful tool for creating more engaging and intuitive human–robot interactions.

Taken together, movement kinematics are a key factor for attribution of specific intentions to others and intentionality in general. In addition, biological motion is also crucial, and in this context, motor experience, visual experience, and emotional processes contribute to whether we perceive other agents as human-like and/or intentional.

4.5 Responsiveness to Social Cues and Predictability

Another interesting aspect of behaviour to consider when evaluating the factors that might affect the attribution of intentionality towards artificial agents is the reactiveness they exhibit to events happening around them. Already in 2007, Terada et al. conducted a study to examine whether individuals attribute some degree of anthropomorphism to artificial agents when they are exposed to reactive movements of non-humanoid robots, such as a chair or a cube with wheels. Participants freely interacted with the artificial agent in an empty room, while its movements were remotely controlled and could either be periodic or reactive. In the reactive condition, the agent's movement followed immediately after the participant's movement, while in the periodic condition, the movement was completely random. After the interaction, participants completed questionnaires to evaluate whether they perceived the agent's movement as intentional or not. The results showed that the physical appearance of the agent (chair vs. cube) did not influence the attribution of intentionality, while the movement profile (periodic vs. reactive) played a significant role in such attributions. The study demonstrated that when the behaviour of an artificial agent is contingent on another event (in this case, human behaviour), participants tend to interpret it as goal-directed, and, thus, intentional. The authors argue that enhancing the sense of intentionality perceived towards a robot is crucial for facilitating effective communication between these agents and humans. Indeed, when a robot appears to be responsive to its environment, it creates in the user the illusion that it can actively perceive and interpret and navigate efficiently that environment, and respond in an appropriate and timely manner to changes. This responsiveness can create a sense of engagement and interaction that closely resembles human-to-human communication (see Brinck & Balkenius, 2018; Bonarini, 2020). The reactivity to the environment seems to be particularly relevant in scenarios where the event causing the response of the robot is generated by the human interacting with it. Indeed, humans are accustomed to communicating with each other through—and be responsive to—verbal and non-verbal cues, such as facial expressions, body language, and gestures. When a robot is capable of perceiving and responding to these cues, it simulates the interactive nature of human-to-human communication. This ability allows the robot to participate actively in the interaction, leading to a more natural and engaging exchange between the human and the robot. Such responsiveness in a robot evokes a sense of engagement from the human side, as in our species we are all prompted to seek reciprocal interactions where our actions and cues are acknowledged and responded to. When a robot exhibits responsiveness, it acknowledges the presence and actions of the human, establishing a reciprocal interaction. This sense of engagement fosters a deeper connection and involvement, making the interaction more meaningful and satisfying for the human. In consequence, this might increase the likelihood of adopting the Intentional Stance towards the robot, as the robot is perceived more 'like-me'. Past research showed that human-to-human communication is characterized by the ability to perceive and interpret subtle cues, understand intentions, and respond accordingly, a sequence of processes that might be elicited, under some circumstances, also in the context of Human–Robot Interaction (Sciutti et al., 2015; Kopp & Kramer, 2021).

Willemse et al. (2018) designed a screen-based experiment to explore the impact of responsiveness of the robot to the signals of the human on social attunement and anthropomorphism. Using eye-tracking, the authors manipulated the robot's behaviour in such a way that it either followed participants' gaze or it looked elsewhere. The authors found that participants (re-) engaged their own attention with the robot faster (after disengagement) when the robot was frequently following their gaze, relative to the condition when it was following less. This shows that robot's responsiveness to human (subtle) behaviour facilitates social attunement. In addition, the results showed that participants evaluated the robot that was more responsive to their own gaze as more human-like and potentially more appealing. In essence, these findings show that our preferences lean towards robots that consistently align their gaze with ours.

In a follow-up study (Willemse & Wykowska, 2019), the authors explored these effects during a real Human–Robot Interaction (i.e. using the physically embodied iCub robot). The authors found similar pattern of results as in their screen-based study. In another experiment, Willemse et al. (2021) confirmed the results of the previous study with a different dependent measure, namely manual responses, instead of saccadic reactions. In addition, in this study, the authors used the InStance Test (Marchesi et al., 2019) to assess whether the responsiveness to the environment modulates attribution of intentionality towards the humanoid. The authors found that participants who were more likely to adopt the Intentional Stance became relatively slow over time when the robot, which mostly did not follow their gaze, followed (occasionally) their object selection. This presumably shows that the robot's unexpected behaviour created some sort of conflict or dissonance, especially in those who adopted the Intentional Stance (thus, were perhaps more inclined to assume that they understand what to expect from the robot).

Reactivity alone implies a certain sensitivity and responsiveness to external stimuli, which might result, during Human–Robot Interaction, in a sense of dynamic adaptiveness of the artificial agent's behaviour to changing circumstances (Tanevska et al., 2020). In human–human interactions, the responsiveness to changes in the environment is, in most cases, predictable to a certain degree (see Dungan et al., 2016; Lustig et al., 2021). It is important to point out that the predictability of human behaviour in response to unexpected events can vary widely depending on individual factors, circumstances, and the nature of the event itself. Indeed, as stated in the previous paragraphs, one of the key features of human behaviour is the variability it displays. However, humans exhibit certain general tendencies in responding to unexpected events or situations, such as the startle reflex or fight-or-flight response (see Burrow, 2009). In a sense, we can say that we 'predict' that a person would react to an event, even though we cannot foretell the exact reaction they will display. For example, if a bear suddenly appears on a mountain path, we might expect all the people on that path to have a reaction, maybe allocating their attentional resources to the potential danger, displaying a certain emotional reaction (such as fear or surprise), trying to walk away (as fast as possible), etc. Of course, some individuals may respond quickly and adaptively to such an unexpected event, drawing on their problem-solving skills, previous experiences, and ability to stay calm under pressure. They may exhibit behaviours such as analysing the situation, taking decisive action, or seeking help and support. On the other hand, some people may initially exhibit a freeze or panic response, temporarily being overwhelmed by the unexpected event. Interestingly, high reactivity may lead to a decreased predictability of behaviour, as it becomes more difficult to anticipate how an individual or system will respond in a given situation (Ajzen, 2011). On the other hand, lower reactivity tends to lead to more stable and predictable behaviour, but it may also limit adaptability and the ability to respond effectively to novel or changing environments (Heiner, 1989). Thus, the degree of reactivity to the environment influences the predictability of behaviour, with higher reactivity yielding greater adaptability but reduced predictability, and lower reactivity resulting in more predictability but potentially limited adaptability. This highlights the need for examining the reactiveness to events happening in the environment in relation to the predictability of the behaviour itself. This aspect might be particularly interesting when considering human–robot interactions, as individuals are less familiar with such technologies and might predict different degrees of reactivity and/or predictability based on their expectations and knowledge about these entities (Horstmann & Kramer, 2020).

Indeed, when robots exhibit predictable behaviours, the human observers might be more likely to attribute intentions to them, considering them as intentional agents (Thellman et al., 2022). In fact, unpredictable behaviours can lead to confusion, distrust, and reduced likelihood of treating the robot as an intentional entity (Papagni & Koeszegi, 2021). This might also depend on the observer's individual need for closure, as discussed in Chapter 3. Predictability in robot's behaviour can increase the user's positive impression by establishing a sense of trust towards the artificial agent (Legaspi et al., 2019). When users can anticipate and understand the intentions behind robot actions, they might feel more in control of the interaction, and this might result in a higher sense of trust towards the robot. Predictability also helps in creating a coherent mental model of the robot's decision-making process, leading to a better understanding of its capabilities and limitations (Matthews et al., 2021). As a result, users could be more likely to rely on the robot and collaborate effectively, ultimately improving the overall HRI experience. On the other hand, the study of Abubshait and Wykowska (2020) showed that extremely repetitive behaviour (which certainly was predictable) reduced the likelihood of adopting the Intentional Stance, which is not surprising, as it lacked the variability that one would expect from a human-like behaviour. A study by Eyssell et al. (2011) suggested that the relationship between predictability and attribution of anthropomorphic traits to a robot might be complex. Their experiment focused on two factors: the anticipation of interaction and the predictability of a robot's behaviour. The researchers hypothesised that already the simple act of anticipating an interaction with a robot can lead to a more pronounced inclination to anthropomorphize the robot. This effect becomes particularly evident when contrasted with situations where the perspective of direct Human–Robot Interaction has not been considered. The results supported these hypotheses, showing that participants' anticipation of a prospective interaction increased participants' tendency to anthropomorphize the robot, especially when the robot's behaviour was unpredictable. However, contrary to expectations, the predictability of the robot's behaviour did not directly lead to increased anthropomorphism. The findings suggested that participants' anticipation of a prospective interaction increased participants' need to understand and control the robot's behaviour, leading to increased anthropomorphism as a way to reduce uncertainty. The study also emphasised the importance of a robot's behavioural characteristics, such as predictability, in facilitating effective HRI. Participants who anticipated interacting with an unpredictable robot were more willing to spend time with it, highlighting that unpredictability did not necessarily reduce acceptance or liking. Based on these results, we can speculate that the predictability of the behaviour displayed by a robot might affect more the joint aspect of HRI than the attribution of anthropomorphic traits to the artificial agent.

In summary, responsiveness of the robot to environmental events, especially if those event are social signals displayed by the observers themselves, increases the attribution of intentionality to the robot, and social attunement. Regarding predictability, there is a complex relationship (and a thin line) between predictable behaviours that increase the 'understandability' (and thus presumably elicit higher degree of adoption of the Intentional Stance), and highly repetitive behaviours that might be perceived as rather mechanical.

4.6 Conclusions

In this chapter, we focused on three main elements that are strictly related to the design of the robot: its appearance, its behaviour/kinematics, and its reactiveness to the environment. The reviewed literature suggests that human-like appearance, human-like behavioural characteristics (temporal parameters, variability, kinematics), as well as biological motion increase anthropomorphism (to no surprise). The chapter reviewed suggestions on how to make a movement appear human-like, through principles of animations. Subsequently, the discussion related to predictability of robot's behaviour concluded with the idea that predictability is a feature that has a more complex impact on adoption of the Intentional Stance, as one needs to find the right balance between endowing robot with predictable (understandable) behaviour while still maintaining the right degree of human-like variability and avoiding too much of rigid repetitiveness that would be perceived rather as a mechanistic and artificial behaviour.

Of note is that in most of the experiments presented here, the main role of the participants was to observe the artificial agents behaving in a more (or less) human-like manner, rather than interacting actively with them. However, interacting with artificial agents, rather than simply observing them, is crucial for forming a more accurate social evaluation of their capabilities (see, for example, Henschel et al., 2020; Heyselaar et al., 2023; Mignone et al., 2023). Indeed, interaction is key for activating mechanisms of social cognition, as emphasized by the second-person neuroscience perspective (see Redcay & Schilbach, 2019). It is important to mention that the 'second-person neuroscience perspective' is an approach in neuroscience that focuses on understanding brain functions during real-time social interactions. Unlike traditional neuroscience which has studied isolated brain activity during responses to stimuli (third-person), the second-person perspective emphasises interactions between individuals. This approach investigates how brains respond during face-to-face interactions, aiming to uncover neural mechanisms of social cognition, empathy, and shared understanding. Studying brain activity in both participants provides insights into social behaviours, emotional sharing, and brain synchronisation, illuminating how our brains engage with others dynamically. Indeed, dynamic interaction provides a more comprehensive understanding of the interaction partner's capabilities, as their responses and actions can be influenced by specific circumstances. Second-person social neuroscience stresses the importance of analysing the brain of two (or more) partners in interaction, in order to understand human social cognition. Similar reasoning can be made in relation to HRI. Indeed, by analysing the human observer in passive observation of a robot, one misses out on abundance of socio-cognitive mechanisms that are activated during interaction. Interacting with artificial agents also allows for multi-modal engagement, involving different senses and modalities. By engaging with the agent, we can assess its visual perception, auditory processing, natural language understanding, and motor skills. This holistic evaluation helps in forming a more comprehensive understanding of the agent's abilities across different domains, which cannot be fully captured through passive observation alone.

In summary, we can claim that interacting with others (including robots) offers a more comprehensive and accurate evaluation of their capabilities compared to passive observation. It provides insights into their contextual understanding, real-time performance, multi-modal abilities, and behavioural adaptation. Through direct interaction, we can gain a deeper understanding of artificial agents' potential and limitations, which should affect adoption of the Intentional Stance to a higher degree than the limited amount of information we can gather by mere observation. The interaction aspect in the context of the Intentional Stance will be further explored in the following chapter of the book.

Interview with Prof. Hiroshi Ishiguro

by Agnieszka Wykowska, Lorenzo Parenti and Ziggy O'Reilly

When I built my first robot, I realised the significance of appearance. That's when I embarked on creating a very human-like android. Fundamentally, my interest wasn't in robots themselves, but in understanding humans better. So I gradually found myself drawn to the creation of highly human-like robots, seeing them as a means to better understand humans.

– Prof. Hiroshi Ishiguro

Prof. Ishiguro with Geminoid™ HI-2. Geminoid™ HI-2 has been developed by Hiroshi Ishiguro Laboratories, Advanced Telecommunications Research Institute International (ATR)

Prof. Hiroshi Ishiguro received a Ph.D. from Osaka University, Japan, in 1991. He is currently Professor of Department of Systems Innovation at Osaka University, Visiting Director of Hiroshi Ishiguro Laboratories at the Advanced Telecommunications Research Institute (ATR), Project Manager of MOONSHOT R&D Project, Thematic Project Producer of EXPO 2025 Osaka, Kansai, Japan, and CEO of AVITA, Inc. His research interests are interactive robotics, avatar, and android science. Geminoid is an avatar android that is a copy of himself. In 2011, he won the Osaka Cultural Award. In 2015, he received the Prize for Science and Technology by the Minister of Education, Culture, Sports, Science and Technology. He was also awarded the Sheikh Mohammed Bin Rashid Al Maktoum Knowledge Award in Dubai in 2015. Tateisi Award in 2020, and honorary doctorate of Aarhus university in 2021.

Summary
On the 31st of January, Professor Agnieszka Wykowska and her two Ph.D. students, Lorenzo Parenti and Ziggy O'Reilly, connected from Italy to speak with Professor Hiroshi Ishiguro in Japan about the role of human-like appearance of robots in evoking the Intentional Stance. The interview covers Prof. Ishiguro's work on robotics, particularly referring to the creation of androids (the Geminoids), implementation of human-likeness in robots, and the use of robot avatars in the near future.

Keywords
Robot appearance, Androids, Robot avatars

Conversation on the Role of Human-Like Appearance in Robots for Human–Robot Interaction

Agnieszka Wykowska Hello Prof. Ishiguro, Thank you very much for taking the time to speak with us. To provide some background, I've led a six-year project funded by the European Research Council focusing on the Intentional Stance towards humanoid robots.

The Intentional Stance, a concept introduced by Daniel Dennett, suggests that when we explain and predict the behavior of other humans, we refer to their mental states such as beliefs, desires, and intentions. The question arises whether, and under what conditions, we also apply this Intentional Stance to humanoid robots. Primarily, we've worked with the iCub robot here at IIT, exploring various factors influencing the likelihood of adopting the Intentional Stance towards it. These factors include human-like behavior and appearance, the level of engagement in interaction, cultural aspects, and background knowledge and exposure to robots.

After 6 years of research, we've decided to summarize our findings in a book, which is where your input becomes extremely valuable to us!

Hiroshi Ishiguro Thank you! Now, if I may ask a question first, I'd like to clarify the distinction between the Intentional Stance and intention itself. What sets Intentional Stance apart from simply having an intention or attitude towards something? Could you elaborate a bit more?

Agnieszka Wykowska Yes, indeed. The Intentional Stance refers to the perspective we, as humans, adopt when explaining and predicting the behaviors of others. Essentially, when we observe a system, an agent, or any entity exhibiting certain behaviors, we apply such stance or strategy to interpret those behaviors.

The concept of intentionality, dating back to Franz Brentano in philosophy, isn't solely about intentions, but encompasses the idea of "aboutness" as a determinant characteristic of all mental states. Intentionality suggests that mental states always have a reference point, or are directed towards something.

Hiroshi Ishiguro I see. So, it's about understanding the origins of this behavior and being able to anticipate what actions the person or agent might take next.

Agnieszka Wykowska Exactly. So usually when we observe a human behaving in a particular manner, we employ what's known as the Intentional Stance.

For instance, if I see a person reaching for a bottle, I might explain their behavior by referring to their desire to drink and their belief that water will quench their thirst. When interacting with a mechanical device like an alarm clock, we apply what's called the Design Stance. We don't attribute mental states to the alarm, but we understand that it was designed to ring at a certain time. There are different stances we adopt: the Intentional Stance towards humans, the Design Stance towards artifacts like cars and coffee machines, and the Physical Stance, which we use to explain, for example, chemical phenomena.

Hiroshi Ishiguro So we could state that the Intentional Stance involves subjective observation and it is based on the "observer perspective", correct?

Agnieszka Wykowska Precisely. The Intentional Stance is essentially our subjective attitude towards a system, whether it's a human, a robot, or anything else. It's like adopting a particular viewpoint or perspective. So, our project aimed to understand when we adopt the Intentional Stance towards humanoid robots.

Hiroshi Ishiguro That's quite fascinating. It seems like in order to adopt Intentional Stance one needs to perceive a system more as an agent rather than just an object. In Japan, for instance, it is culturally very common to attribute or see a soul or agency in something that it is not necessarily human. So, what's your question?

Agnieszka Wykowska Well, I have a more personal question for you. I'd like to know when your interest in robotics, particularly in androids, began and where the idea of creating such a human-like android originated.

Hiroshi Ishiguro Ah, I see. Well, to be honest, I wasn't initially interested in robots.

> ***My curiosity always centered around understanding myself and other humans. I remember back when I was around 12 years old, in the fifth year of elementary school, adults would often tell me to consider what others were thinking. But I found it incredibly challenging. How does one think about another person's mind? It seemed like an impossible task. The concept of the mind was abstract to me; I couldn't visualise it concretely.***

So, I became increasingly curious about the mind, thinking, and even the fundamental aspects of being. Eventually, I pursued studies in computer science, which naturally led me to delve into AI. As I delved deeper into AI, I realized the importance of embodiment. I felt that to truly understand humans, I needed to give AI a body. Hence, I ventured into robotics. When I built my first robot, I realized the significance of appearance. That's when I embarked on creating a very human-like android. Fundamentally, my interest wasn't in robots themselves but in understanding humans better. So I gradually found myself drawn to the creation of highly human-like robots, seeing them as a means to better understand humans.

Agnieszka Wykowska This is incredibly fascinating. It seems like we're aligned in our approach. Building robots, and using them to study the human and social interaction.

Hiroshi Ishiguro Yes, I began the interactive robot project around 2000 and have spent over 24 years in this field. However, human beings are incredibly complex, and I'm not sure if I can definitely say that building robots has brought me closer to understanding humans. Yet, creating androids allows me to grapple with this complexity more directly. For instance, my Geminoid now speaks on my behalf. It's incredibly convenient. To give an example, I no longer need to personally greet visitors in the lab in case I am unable to travel or attend. My Geminoid is capable of managing such tasks and engaging in social interactions in my place. It closely resembles me, and I've integrated within it [the robot] over ten books authored by me, along with transcripts of various interviews I've conducted. This enables it to effectively speak on my behalf, utilizing advanced language models.

Agnieszka Wykowska It's truly fascinating to consider the possibilities for the future. Thank you so much. I'll now pass the conversation over to Lorenzo, who has prepared some questions from his perspective.

Lorenzo Parenti Thank you very much Professor Ishiguro, for accepting to do this interview. It's an honor to have the possibility to ask you some questions about your experience. Firstly, I would like to ask you about the importance you've placed on human-like appearance in robots, stating that it can enhance the user's perception of the robot's presence. Could you elaborate on how the Intentional Stance, or ascribing intentions and mental states to robots, can play a role in this perceived presence of those robots and robotic agents?

Hiroshi Ishiguro Certainly. I'd like to share some insights from our recent studies. Human-like appearance is indeed crucial for eliciting the Intentional Stance quickly. Without a highly human-like appearance, we can still sense strong intentionality from the robot. In the installation we recently developed, we do not use a very human-like robot; instead, we opt for a mechanical-looking one. However, this robot can convey human-like gestures and do eye contact. The system captures visitors' movements, and responds to them non-verbally. Through these signals, visitors can still perceive a human-like quality.

> ***While giving a robot a human-like appearance is beneficial for sensing intentionality, sometimes appearance is not as crucial as interactive behaviors. Though having human-like features such as arms, legs, and eyes is preferred, what truly matters is the interaction itself. Appearance serves as a trigger, but it is not as dynamic as gestures or other non-verbal behaviors.***

Appearance can influence these dynamics, but if the behavior isn't up to par, it can lead to an uncanny feeling. So, while appearance matters, interactional behavior holds more significance in establishing a sense of intentionality.

Lorenzo Parenti Thank you for that insightful explanation. You've expressed the possibility of building androids that are nearly indistinguishable from humans, at least during brief encounters. How do you think the Intentional Stance contributes to this sense of indistinguishability between humans and androids, and what considerations should be taken into account when striving for that resemblance?

Hiroshi Ishiguro For brief interactions, a human-like appearance is crucial for sensing intentionality quickly. However, appearance becomes less critical for longer interactions, where behaviours and conversations take precedence. Interaction itself becomes more important. Appearance acts as a trigger, but it's not a dynamic factor. We need dynamic interactions to fully perceive intentionality in a long run. While appearance can influence these dynamics, it's not the sole determining factor.

Therefore, while it's preferable to have a very human-like appearance, the quality of behavior and interaction matter more. For instance, even if a robot lacks certain human-like features, such as arms or legs (Fig. 4.2), it can still convey strong intentionality through its behaviors. Ultimately, the interaction itself is what matters most in establishing a sense of intentionality.

Ziggy O'Reilly You mentioned earlier the cultural differences in perceiving intentionality in robots, referring to the Japanese culture's tendency to attribute a soul to various artifacts. How do you navigate these cultural differences when designing robots? Do you consider the specific culture you're designing the robot for?

Hiroshi Ishiguro Primarily, I conduct field experiments and tests in Japan, although I've recently started collaborating with researchers in Barcelona and Dubai. Japan generally has a positive attitude toward robots and my focus has mainly been on Japanese culture. However, I recognise the importance of studying cross-cultural issues, especially as I aim to introduce interactive robots to foreign countries. I'm particularly interested in how different cultures perceive robot appearance and behavior. For instance, in my computer graphics business, I've noticed that Japanese companies prefer cartoonish avatars, while foreign companies prefer realistic avatars. These differences in preference reflect broader cultural variations in perceptions of appearance and robot aesthetics.

Ziggy O'Reilly Thank you for sharing those insights. One last question regarding this topic; how do you perceive the use of [robot] avatars in mediated communication? Are you comfortable with using avatars for interactions, especially when discussing private or personal matters?

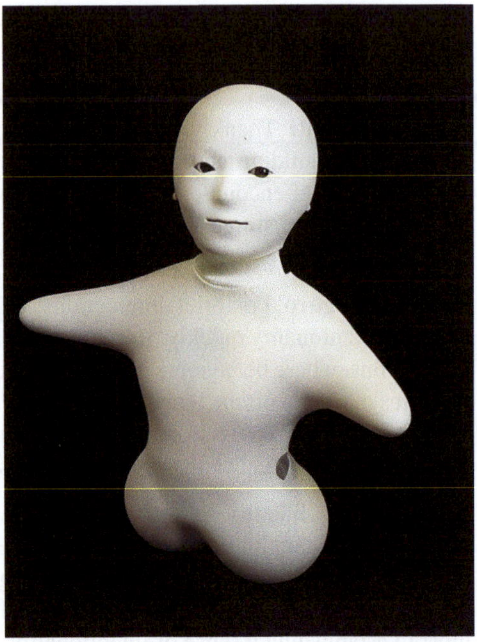

Fig. 4.2 A Telenoid™ robot which has been developed by Osaka University and Hiroshi Ishiguro Laboratories, Advanced Telecommunications Research Institute International (ATR)

Hiroshi Ishiguro Thank you for asking. It depends on the context. For short, transactional interactions, avatars can be convenient and less intimidating than embodied robots or androids. However, for deeper, longer-term relationships, I prefer direct and embodied communication. The use of avatars for sensitive topics like privacy can be acceptable if done correctly and securely [...]

Sometimes people tend to be reluctant to discuss their privacy with other humans. But with avatars, I would expect that they can open up more easily. They can discuss privacy issues without hesitation. This enables avatar operators to offer better services.

But of course, for long-term interactions and relationships, we'll likely opt for face-to-face interactions, in our own bodies. However, for instantaneous conversations, avatars work very well. For instance, let's take the situation of purchasing something at a shop. We prefer not to form deep relationships with shopkeepers, right? In such cases, it's better to use avatars. We don't hesitate to communicate with avatars. They're a type of robot, not real humans, so we don't need to worry that much about privacy issues or anything like that. I'm still not completely sure how people would accept this kind of service. So that's one of my new challenges now.

Ziggy O'Reilly I see. Thank you for your time and for sharing your perspective with us.

Hiroshi Ishiguro Well, I'm afraid I have another meeting to attend now, but I appreciate the opportunity to discuss these topics with you and hope to visit IIT and Italy soon.

References

Abubshait, A., & Wiese, E. (2017). You look human, but act like a machine: Agent appearance and behavior modulate different aspects of human–robot interaction. *Frontiers in Psychology, 8*. https://doi.org/10.3389/fpsyg.2017.01393

Abubshait, A., & Wykowska, A. (2020). Repetitive robot behavior impacts perception of intentionality and gaze-related attentional orienting. *Frontiers in Robotics and AI*, 7. https://doi.org/10.3389/frobt.2020.565825

Abubshait, A., Momen, A., & Wiese, E. (2020). Pre-exposure to ambiguous faces modulates top-down control of attentional orienting to counterpredictive gaze cues. *Frontiers in Psychology, 2234*, 10.31234/osf.io/7ezpb.

Ajzen, I. (2011). The theory of planned behaviour: Reactions and reflections. *Psychology & Health, 26*(9), 1113–1127. https://doi.org/10.1080/08870446.2011.613995

Atkinson, A. P., Vuong, Q. C., & Smithson, H. E. (2012). Modulation of the face- and body-selective visual regions by the motion and emotion of point-light face and body stimuli. *NeuroImage, 59*(2), 1700–1712. https://doi.org/10.1016/j.neuroimage.2011.08.073

Axelrod, L., & Hone, K. (2005). *E-motional advantage*. In Proceedings of ACM CHI 2005 conference on human factors in computing systems. https://doi.org/10.1145/1056808.1056874.

Bartneck, C., & Forlizzi, J. (2004). *Shaping human-robot interaction*. In Proceedings of ACM CHI 2004 conference on human factors in computing systems. https://doi.org/10.1145/985921.986205.

Bisio, A., Sciutti, A., Nori, F., Metta, G., Fadiga, L., Sandini, G., & Pozzo, T. (2014). Motor contagion during human-human and human-robot interaction. *PLoS One, 9*(8), e106172. https://doi.org/10.1371/journal.pone.0106172

Bonarini, A. (2020). Communication in human-robot interaction. *Current Robotics Reports, 1*(4), 279–285. https://doi.org/10.1007/s43154-020-00026-1

Bonini, L., Rotunno, C., Arcuri, E., & Gallese, V. (2022). Mirror neurons 30 years later: Implications and applications. *Trends in Cognitive Sciences, 26*(9), 767–781. https://doi.org/10.1016/j.tics.2022.06.003

Brinck, I., & Balkenius, C. (2018). Mutual recognition in human-robot interaction: A deflationary account. *Philosophy & Technology, 33*(1), 53–70. https://doi.org/10.1007/s13347-018-0339-x

Burrow, S. (2009). 6 bodily limits to autonomy: Emotion, attitude, and self-defense. *Embodiment and Agency*, 126–142. https://doi.org/10.1515/9780271079509-008

Cavallo, A., Koul, A., Ansuini, C., Capozzi, F., & Becchio, C. (2016). Decoding intentions from movement kinematics. *Scientific Reports, 6*(1). https://doi.org/10.1038/srep37036

Chaminade, T., Hodgins, J., & Kawato, M. (2007). Anthropomorphism influences perception of computer-animated characters' actions. *Social Cognitive and Affective Neuroscience*, 2(3), 206–216. https://doi.org/10.1093/scan/nsm017

Chaminade, T., Zecca, M., Blakemore, S.-J., Takanishi, A., Frith, C. D., Micera, S., Dario, P., Rizzolatti, G., Gallese, V., & Umiltà, M. A. (2010). Brain response to a humanoid robot in areas implicated in the perception of human emotional gestures. *PLoS One, 5*(7), e11577. https://doi.org/10.1371/journal.pone.0011577

Cheetham, M. (2017). Editorial: The uncanny valley hypothesis and beyond. *Frontiers in Psychology, 8*. https://doi.org/10.3389/fpsyg.2017.01738

Ciardo, F., De Tommaso, D., & Wykowska, A. (2022). Human-like behavioral variability blurs the distinction between a human and a machine in a nonverbal turing test. *Science Robotics, 7*(68). https://doi.org/10.1126/scirobotics.abo1241

Cross, E. S., Liepelt, R., de Hamilton, A. F., Parkinson, J., Ramsey, R., Stadler, W., & Prinz, W. (2011). Robotic movement preferentially engages the action observation network. *Human Brain Mapping, 33*(9), 2238–2254. Portico. https://doi.org/10.1002/hbm.21361

Desimone, R., & Duncan, J. (1995). Neural mechanisms of selective visual attention. *Annual Review of Neuroscience, 18*(1), 193–222. https://doi.org/10.1146/annurev.ne.18.030195.001205

di Pellegrino, G., Fadiga, L., Fogassi, L., Gallese, V., & Rizzolatti, G. (1992). Understanding motor events: A neurophysiological study. *Experimental Brain Research, 91*(1), 176–180. https://doi.org/10.1007/bf00230027

Duffy, B. R. (2003). Anthropomorphism and the social robot. *Robotics and Autonomous Systems, 42*(3–4), 177–190. https://doi.org/10.1016/s0921-8890(02)00374-3

Dungan, J. A., Stepanovic, M., & Young, L. (2016). Theory of mind for processing unexpected events across contexts. *Social Cognitive and Affective Neuroscience, 11*(8), 1183–1192. https://doi.org/10.1093/scan/nsw032

Epley, N., Waytz, A., & Cacioppo, J. T. (2007). On seeing human: A three-factor theory of anthropomorphism. *Psychological Review, 114*(4), 864–886. https://doi.org/10.1037/0033-295x.114.4.864

Eyssel, F., Kuchenbrandt, D., & Bobinger, S. (2011). *Effects of anticipated human-robot interaction and predictability of robot behavior on perceptions of anthropomorphism.* In Proceedings of the 6th international conference on human-robot interaction (pp. 61–68). https://doi.org/10.1145/1957656.1957673.

Ferrari, F., Paladino, M. P., & Jetten, J. (2016). Blurring human–machine distinctions: Anthropomorphic appearance in social robots as a threat to human distinctiveness. *International Journal of Social Robotics, 8*(2), 287–302. https://doi.org/10.1007/s12369-016-0338-y

Ferrey, A. E., Burleigh, T. J., & Fenske, M. J. (2015). Stimulus-category competition, inhibition, and affective devaluation: A novel account of the uncanny valley. *Frontiers in Psychology, 6*. https://doi.org/10.3389/fpsyg.2015.00249

Fink, J. (2012). *Anthropomorphism and human likeness in the design of robots and human-robot interaction.* In Social robotics: 4th international conference, ICSR 2012, Chengdu, China, October 29–31, 2012. Proceedings 4 (pp. 199–208). Springer. https://doi.org/10.1007/978-3-642-34103-8_20.

Freedman, E. G., & Sparks, D. L. (2000). Coordination of the eyes and head: Movement kinematics. *Experimental Brain Research, 131*(1), 22–32. https://doi.org/10.1007/s002219900296

Frith, C. D., & Frith, U. (2006). The neural basis of mentalizing. *Neuron, 50*(4), 531–534. https://doi.org/10.1016/j.neuron.2006.05.001

Gallese, V., Fadiga, L., Fogassi, L., & Rizzolatti, G. (1996). Action recognition in the premotor cortex. *Brain, 119*(2), 593–609. https://doi.org/10.1093/brain/119.2.593

Gazzola, V., Rizzolatti, G., Wicker, B., & Keysers, C. (2007). The anthropomorphic brain: The mirror neuron system responds to human and robotic actions. *NeuroImage, 35*, 1674–1684. https://doi.org/10.1016/j.neuroimage.2007.02.003

Ghiglino, D., & Wykowska, A. (2020). When robots (pretend to) think. *Artificial Intelligence*, 49–74. https://doi.org/10.30965/9783957437488_006

Ghiglino, D., Willemse, C., Tommaso, D. D., Bossi, F., & Wykowska, A. (2020a). At first sight: Robots' subtle eye movement parameters affect human attentional engagement, spontaneous attunement and perceived human-likeness. *Paladyn, Journal of Behavioral Robotics, 11*(1), 31–39. https://doi.org/10.1515/pjbr-2020-0004

Ghiglino, D., De Tommaso, D., Willemse, C., Marchesi, S., & Wykowska, A. (2020b). Can I get your (robot) attention? Human sensitivity to subtle hints of human-likeness in a humanoid robot's behavior. *In Cogsci 2020.*

Ghiglino, D., Willemse, C., De Tommaso, D., & Wykowska, A. (2021). Mind the eyes: Artificial agents' eye movements modulate attentional engagement and anthropomorphic attribution. *Frontiers in Robotics and AI*, 8. https://doi.org/10.3389/frobt.2021.642796

Gielniak, M. J., Liu, C. K., & Thomaz, A. L. (2013). Generating human-like motion for robots. *The International Journal of Robotics Research, 32*(11), 1275–1301. https://doi.org/10.1177/0278364913490533

Hegel, F., Krach, S., Kircher, T., Wrede, B., & Sagerer, G. (2008). *Understanding social robots: A user study on anthropomorphism.* In RO-MAN 2008—The 17th IEEE international symposium on robot and human interactive communication (pp. 574–579). https://doi.org/10.1109/roman.2008.4600728.

Heider, F., & Simmel, M. (1944). An experimental study of apparent behavior. *The American Journal of Psychology, 57*(2), 243. https://doi.org/10.2307/1416950

Heiner, R. A. (1989). The origin of predictable dynamic behavior. *Journal of Economic Behavior & Organization, 12*(2), 233–257. https://doi.org/10.1016/0167-2681(89)90057-7

Henschel, A., Hortensius, R., & Cross, E. S. (2020). Social cognition in the age of human–robot interaction. *Trends in Neurosciences, 43*(6), 373–384. https://doi.org/10.1016/j.tins.2020.03.013

Heyselaar, E., Caruana, N., Shin, M., Schilbach, L., & Cross, E. S. (2023). Do we really interact with artificial agents as if they are human? *Frontiers in Virtual Reality, 4*, 1201385. https://doi.org/10.3389/frvir.2023.1201385

Horstmann, A. C., & Krämer, N. C. (2020). Expectations vs. actual behavior of a social robot: An experimental investigation of the effects of a social robot's interaction skill level and its expected future role on people's evaluations. *PLoS One, 15*(8), e0238133. https://doi.org/10.1371/journal.pone.0238133

Johansson, G. (1973). Visual perception of biological motion and a model for its analysis. *Perception & Psychophysics, 14*(2), 201–211. https://doi.org/10.3758/bf03212378

Kaplan, F. (2004). Who is afraid of the humanoid? Investigatingcultural differences in the acceptance of robots. *International Journal of Humanoid Robotics, 01*(03), 465–480. https://doi.org/10.1142/s0219843604000289

Kilner, J. M., Paulignan, Y., & Blakemore, S. J. (2003). An interference effect of observed biological movement on action. *Current Biology, 13*, 522–525. https://doi.org/10.1016/S0960-9822(03)00165-9

Kopp, S., & Krämer, N. (2021). Revisiting human-agent communication: The importance of joint co-construction and understanding mental states. *Frontiers in Psychology, 12*. https://doi.org/10.3389/fpsyg.2021.580955

Koul, A., Soriano, M., Tversky, B., Becchio, C., & Cavallo, A. (2019). The kinematics that you do not expect: Integrating prior information and kinematics to understand intentions. *Cognition, 182*, 213–219. https://doi.org/10.1016/j.cognition.2018.10.006

Kupferberg, A., Huber, M., Helfer, B., Lenz, C., Knoll, A., & Glasauer, S. (2012). Moving just like you: Motor interference depends on similar motility of agent and observer. *PLoS One, 7*(6), e39637. https://doi.org/10.1371/journal.pone.0039637

Lammers, S., Bente, G., Tepest, R., Jording, M., Roth, D., & Vogeley, K. (2019). Introducing ACASS: An annotated character animation stimulus set for controlled (e)motion perception studies. *Frontiers in Robotics and AI*, 6. https://doi.org/10.3389/frobt.2019.00094

Lasseter, J. (1998). Principles of traditional animation applied to 3D computer animation. *Seminal Graphics*, 263–272. https://doi.org/10.1145/280811.281005

Legaspi, R., He, Z., & Toyoizumi, T. (2019). Synthetic agency: Sense of agency in artificial intelligence. *Current Opinion in Behavioral Sciences, 29*, 84–90. https://doi.org/10.1016/j.cobeha.2019.04.004

Looser, C. E., & Wheatley, T. (2010). The tipping point of animacy. *Psychological Science, 21*(12), 1854–1862. https://doi.org/10.1177/0956797610388044

Lustig, C., Esser, S., & Haider, H. (2021). The interplay between unexpected events and behavior in the development of explicit knowledge in implicit sequence learning. *Psychological Research, 86*(7), 2225–2238. https://doi.org/10.1007/s00426-021-01630-2

MacDorman, K. F., Vasudevan, S. K., & Ho, C.-C. (2008). Does Japan really have robot mania? Comparing attitudes by implicit and explicit measures. *AI & SOCIETY, 23*(4), 485–510. https://doi.org/10.1007/s00146-008-0181-2

Marchesi, S., Ghiglino, D., Ciardo, F., Perez-Osorio, J., Baykara, E., & Wykowska, A. (2019). Do we adopt the intentional stance toward humanoid robots? *Frontiers in Psychology, 10*. https://doi.org/10.3389/fpsyg.2019.00450

Mather, G., & Murdoch, L. (1994). Gender discrimination in biological motion displays based on dynamic cues. *Proceedings of the Royal Society of London. Series B: Biological Sciences, 258*(1353), 273–279. https://doi.org/10.1098/rspb.1994.0173

Matthews, G., Panganiban, A. R., Lin, J., Long, M., & Schwing, M. (2021). Super-machines or sub-humans: Mental models and trust in intelligent autonomous systems. *Trust in Human-Robot Interaction, 59–82*. https://doi.org/10.1016/b978-0-12-819472-0.00003-4

Michotte, A. (1963). *The perception of causality* (T. R. Miles & E. Miles, Trans.). Basic Books. (Original work published 1946). https://doi.org/10.1093/bjps/xv.59.254.

Mignone, G., Parziale, A., Ferrentino, E., Marcelli, A., & Chiacchio, P. (2023). Observation vs. interaction in the recognition of human-like movements. *Frontiers in Robotics and AI, 10*. https://doi.org/10.3389/frobt.2023.1112986

Mori, M. (1970). The uncanny valley: The original essay by Masahiro Mori. *IEEE Spectrum, 6*. https://doi.org/10.1109/MRA.2012.2192811

Mori, M., MacDorman, K., & Kageki, N. (2012). The uncanny valley [from the field]. *IEEE Robotics & Automation Magazine, 19*(2), 98–100. https://doi.org/10.1109/mra.2012.2192811

Padois, V., Ivaldi, S., Babič, J., Mistry, M., Peters, J., & Nori, F. (2017). Whole-body multi-contact motion in humans and humanoids: Advances of the CoDyCo European project. *Robotics and Autonomous Systems, 90*, 97–117. https://doi.org/10.1016/j.robot.2016.08.017

Papagni, G., & Koeszegi, S. (2021). A pragmatic approach to the intentional stance semantic, empirical and ethical considerations for the design of artificial agents. *Minds and Machines, 31*(4), 505–534. https://doi.org/10.1007/s11023-021-09567-6

Parenti, L., Marchesi, S., Belkaid, M., & Wykowska, A. (2021). *Exposure to robotic virtual agent affects adoption of intentional stance*. In Proceedings of the 9th international conference on human-agent interaction. https://doi.org/10.1145/3472307.3484667.

Parenti, L., Marchesi, S., Belkaid, M., & Wykowska, A. (2022). Attributing intentionality to artificial agents: Exposure versus interactive scenarios. *Lecture Notes in Computer Science, 347–356*. https://doi.org/10.1007/978-3-031-24667-8_31

Perez-Osorio, J., & Wykowska, A. (2019). Adopting the intentional stance towards humanoid robots. In *Springer tracts in advanced robotics* (pp. 119–136). https://doi.org/10.1007/978-3-030-17974-8_10

Redcay, E., & Schilbach, L. (2019). Using second-person neuroscience to elucidate the mechanisms of social interaction. *Nature Reviews Neuroscience, 20*(8), 495–505. https://doi.org/10.1038/s41583-019-0179-4

Reuten, A., van Dam, M., & Naber, M. (2018). Pupillary responses to robotic and human emotions: The uncanny valley and media equation confirmed. *Frontiers in Psychology, 9*. https://doi.org/10.3389/fpsyg.2018.00774

Riek, L. D., Rabinowitch, T.-C., Chakrabarti, B., & Robinson, P. (2009). *How anthropomorphism affects empathy toward robots*. In Proceedings of the 4th ACM/IEEE international conference on human robot interaction. https://doi.org/10.1145/1514095.1514158.

Rosenthal-von der Pütten, A. M., & Krämer, N. C. (2014). How design characteristics of robots determine evaluation and uncanny valley related responses. *Computers in Human Behavior, 36*, 422–439. https://doi.org/10.1016/j.chb.2014.03.066

Sciutti, A., Ansuini, C., Becchio, C., & Sandini, G. (2015). Investigating the ability to read others' intentions using humanoid robots. *Frontiers in Psychology, 6*. https://doi.org/10.3389/fpsyg.2015.01362

Shiffrar, M., & Thomas, J. P. (2013). Beyond the scientific objectification of the human body. In *Social perception: Detection and interpretation of animacy, agency, and intention* (pp. 83–108). https://doi.org/10.7551/mitpress/9780262019279.003.0005

Tanevska, A., Rea, F., Sandini, G., Cañamero, L., & Sciutti, A. (2020). A socially adaptable framework for human-robot interaction. *Frontiers in Robotics and AI, 7*. https://doi.org/10.3389/frobt.2020.00121

Terada, K., Shamoto, T., Ito, A., & Mei, H. (2007). *Reactive movements of non-humanoid robots cause intention attribution in humans*. In 2007 IEEE/RSJ international conference on intelligent robots and systems. https://doi.org/10.1109/iros.2007.4399429.

Thellman, S., de Graaf, M., & Ziemke, T. (2022). Mental state attribution to robots: A systematic review of conceptions, methods, and findings. *ACM Transactions on Human-Robot Interaction, 11*(4), 1–51. https://doi.org/10.1145/3526112

Thomas, F., & Johnston, O. (1981). "The illusion of life": Nature in the animated Disney CALEB STEINDAM. *Disney, Culture, and Curriculum*, 117–132. https://doi.org/10.4324/9781315661599-17

Vaitonytė, J., Alimardani, M., & Louwerse, M. M. (2023). Scoping review of the neural evidence on the uncanny valley. *Computers in Human Behavior Reports, 9*, 100263. https://doi.org/10.1016/j.chbr.2022.100263

Varella, M. A. C. (2018). The biology and evolution of the three psychological tendencies to anthropomorphize biology and evolution. *Frontiers in Psychology, 9*. https://doi.org/10.3389/fpsyg.2018.01839

Wiese, E., Metta, G., & Wykowska, A. (2017). Robots as intentional agents: Using neuroscientific methods to make robots appear more social. *Frontiers in Psychology, 8*. https://doi.org/10.3389/fpsyg.2017.01663

Wiese, E., Mandell, A., Shaw, T., & Smith, M. (2019). Implicit mind perception alters vigilance performance because of cognitive conflict processing. *Journal of Experimental Psychology: Applied, 25*(1), 25–40. https://doi.org/10.1037/xap0000186

Willemse, C., & Wykowska, A. (2019). In natural interaction with embodied robots, we prefer it when they follow our gaze: A gaze-contingent mobile eyetracking study. *Philosophical Transactions of the Royal Society B: Biological Sciences, 374*(1771), 20180036. https://doi.org/10.1098/rstb.2018.0036

Willemse, C., Marchesi, S., & Wykowska, A. (2018). Robot faces that follow gaze facilitate attentional engagement and increase their likeability. *Frontiers in Psychology, 9*. https://doi.org/10.3389/fpsyg.2018.00070

Willemse, C., Abubshait, A., & Wykowska, A. (2021). Motor behaviour mimics the gaze response in establishing joint attention, but is moderated by individual differences in adopting the intentional stance towards a robot avatar. *Visual Cognition, 30*(1–2), 42–53. https://doi.org/10.1080/13506285.2021.1994494

Yamane, K., & Murai, A. (2018). A comparative study between humans and humanoid robots. *Humanoid Robotics: A Reference*, 873–892. https://doi.org/10.1007/978-94-007-6046-2_7

Złotowski, J., Sumioka, H., Nishio, S., Glas, D. F., Bartneck, C., & Ishiguro, H. (2016). Appearance of a robot affects the impact of its behaviour on perceived trustworthiness and empathy. *Paladyn, Journal of Behavioral Robotics, 7*(1). https://doi.org/10.1515/pjbr-2016-0005

Chapter 5
Factors Influencing the Adoption of the Intentional Stance: The Interaction

Abdulaziz Abubshait, Marwen Belkaid, Kyveli Kompatsiari, Uma Prashant Navare, and Jairo Perez-Osorio

The Encyclopaedia Galactica defines a robot as a mechanical apparatus designed to do the work of a man.
The marketing division of the Sirius Cybernetics Corporation defines a robot as 'Your Plastic Pal Who's Fun to Be With'.

– Douglas Adams, The Hitchhiker's Guide to the Galaxy.

5.1 Introduction

In the previous chapters, we discussed the robot design on the one hand: how engineers and robot manufacturers often consider endowing robots with human-like appearance, communication abilities, movement, and adaptability to increase how

A. Abubshait (✉)
Italian Institute of Technology, Social Cognition in Human-Robot Interaction Unit, Genova, Italy

M. Belkaid · U. P. Navare
Italian Institute of Technology, Social Cognition in Human-Robot Interaction Unit, Genova, Italy

ETIS UMR 8051, CY Cergy Paris Université, ENSEA, CNRS, Cergy, France
e-mail: marwen.belkaid@ensea.fr

K. Kompatsiari
Italian Institute of Technology, Social Cognition in Human-Robot Interaction Unit, Genova, Italy

Department of Applied Mathematics and Computer Science, Cognitive Systems, Technical University of Denmark, Kongens Lyngby, Denmark

J. Perez-Osorio
Italian Institute of Technology, Social Cognition in Human-Robot Interaction Unit, Genova, Italy

Cognitive Psychology & Ergonomics, Technische Universität Berlin, Berlin, Germany

A. Wykowska (ed.), *Intentional Stance Towards Humanoid Robots*,
https://doi.org/10.1007/978-3-031-65483-1_5

engaging these robots can be; and the human, on the other hand: how specific mindset, expertise, or experience with technology affect the way people approach robots. However, another key component affecting the adoption of the Intentional Stance is ***the interaction itself***.

In this chapter, we delve into the interaction theme: the complex interplay between humans and humanoid robots. We examine how the context of interaction itself can increase or decrease the likelihood that a robot is being perceived as intentional. It is important to note here that we focus only on short-term interactions, as we limit our discussion only to controlled lab experiments.

We aim to underline the breadth of the term 'interaction', which encompasses a diverse range of activities, and illustrate what makes each type of interaction unique and how these different interaction types affect perceptions of a robot's intentionality, again, as measured by the InStance Test (Marchesi et al., 2019) or adjacent tools.

When examining Human–Robot Interaction closely, we create a taxonomy of the numerous variations of the 'interaction contexts' and find that we are able to place these different types of interaction along a spectrum of engagement. This spectrum spans a continuum of shared tasks, shared objectives, and shared experiences (or at least a belief in shared experiences). This gradient of interaction contexts comprises intermediary stages, each characterized by distinct degrees of levels of engagement within the realm of interaction. According to these degrees or 'levels' of interaction contexts, we group also experimental protocols into three major groups. The breakdown of these protocols is as follows:

1. Protocols involving ***Sharing Environmental Stimuli***: This category signifies interaction that goes beyond mere observation. It encompasses instances where the robot responds to and shares the environmental stimuli with the human, forging a bridge between passive observation and more sophisticated forms of interaction.
2. Protocols involving ***Belief in Sharing Experience, Without Joint Goals***: At this stage, interaction involves the exchange of experiences without a common purpose or objective. The human believes that they and the robot share their respective experiences without working towards a common goal.
3. Protocols involving ***Belief in Sharing Experience, with Joint Goals***: The pinnacle of interaction is achieved when both the human and the robot collaborate, whilst the human beliefs in shared experience. In addition, the human and the robot actually share common objectives or goals. This dynamic involves a deeper level of engagement, where both entities synchronize their actions to achieve shared outcomes.

In essence, this chapter explores the multifaceted nature of Human–Robot Interaction, dissecting it into a gradient of engagement levels and categorizing it into various experimental protocols, while examining how each level affects the attributions and perceptions of intentionality.

5.2 Sharing Environmental Stimuli

In the spectrum of experimental protocols addressing the issue of Human–Robot Interaction and its impact on the adoption of the Intentional Stance, the next level of 'interactivity', after pure observation protocols (Chapter 4), comprises protocols in which the human and the robot react to similar environmental stimuli. This can be classified as 'joint attention', meaning that two interaction partners attend to the same location or event in the environment. In the case of joint attention, the human interactor can either be responding to robot's behaviour (e.g. the robot looks towards one location, and the human attends there too) or the human can be the initiator of joint attention (e.g. the robot responds to the human's directional gaze by looking in the same location as the human does). We will discuss the circumstances in which this level of interaction between the human and robot affected the likelihood of adopting the Intentional Stance towards the robot and those in which it did not.

Abubshait and Wykowska (2020) explored how a well-known psychological phenomenon, known as the gaze-cueing effect (which is a manifestation of joint attention), correlates with adopting the Intentional Stance. A gaze-cueing protocol is an adaptation of the classical Posner paradigm (Posner, 1980) and typically presents a simple face on a computer screen that directs the participant's attention to a location where a stimulus (e.g., a letter) would appear on the screen (Friesen & Kingstone, 1998). Participants then respond to this stimulus by indicating its identity (e.g., whether it is a letter 'F' or 'T') or its location (e.g., left vs. right). The typical pattern of results shows that performance (reaction times, accuracy) is better for letters which appear at the gazed-at location, relative to other locations. This is interpreted as an effect of allocation of attention in line with the direction of the gaze, that is, engagement in joint attention. In the study of Abubshait and Wykowska (2020), participants were seated in front of a humanoid robot, instead of a computer screen. The robot would shift its gaze to one of two screens placed on its sides such setup embodied the gaze shift that typically occurs in the screen-based experiments. Participants would then respond to the target letter using a keyboard, as they would do in the computer-based version of the protocol. Similar experimental setups can be seen in studies by Kompatsiari and colleagues (Kompatsiari et al., 2017, 2018, 2021, 2022). In their study, Abubshait and Wykowska set out to investigate how the duration of exposure to a humanoid robot's repetitive behaviour (gazing to the left or right with a fixed timing and movement trajectories) would affect observers' likelihood to adopt the Intentional Stance to the robot. Their findings showed that shorter interactions (smaller number of trials in the gaze-cueing protocol) led to a higher likelihood that participants would adopt the Intentional Stance to the robot. On the contrary, longer exposures to the robot's repetitive behaviour (larger number of trials in the gaze-cueing protocol) decreased this. Interestingly, however, as exposure duration increased, participants were more likely to exhibit larger joint attention effects (gaze cueing). Thus, an interesting dilemma emerged: increasing the number of experimental trials (and thereby duration of interaction) increased engagement in joint attention (whch is one of the mechanisms of social attunement), but reduced the likelihood of adopting the Intentional Stance. This interesting dissociation might be considered in future studies.

In a series of studies similar to that of Abubshait and Wykowska (2020), Kompatsiari et al. (2018, 2019a, b, 2021, 2022) examined the impact of eye contact with a humanoid robot on gaze cueing and social attunement in general. In several experiments, Kompatsiari and colleagues repeated the same procedure: participants sat across from the iCub robot in a gaze-cueing setup that was aforementioned. In addition to the gaze-cueing procedure (directional gaze of the iCub robot), the researchers manipulated gaze contact (direct or averted gaze), prior to the directional cue. Related to the perception of robot's gaze, participants rated the robot as more anthropomorphic and likeable when it engaged in eye contact, compared to averted gaze (on the respective scales of Godspeed questionnaire) (Bartneck et al., 2009). Interestingly, most participants who perceived the robot as human-like based their reasoning on the establishment of eye contact (Kompatsiari et al., 2019b). Moreover, in their experiments (Kompatsiari et al., 2018, 2019a, b, 2021, 2022), the authors collected behavioural responses (reaction times to target letters), as well as eye-tracking data (Kompatsiari et al., 2019a) or EEG data (Kompatsiari et al., 2021). Notably, after each block of gaze-cueing trials, participants were asked to rate how engaging the robot's eye contact felt to them. Across all experiments, the researchers observed that eye contact with the robot elicited higher engagement, relative to averted gaze: this was manifested through higher subjective ratings, neural indices of engagement (higher desynchronization in the alpha frequency), as well as longer fixation durations on the robot face. Engagement due to the social signal of a communicative gaze might be the driving force that facilitated subsequent joint attention, that is, higher gaze-cueing effects in the eye contact condition, compared to averted gaze (Kompatsiari et al., 2018, 2019a, 2022). The potential social origin of the effect of a communicative gaze on joint attention is suggested by a follow-up study where the modulatory effect of eye contact/averted gaze on attentional orienting was eliminated by disrupting areas of the brain related to social cognition (brain stimulation on dorso-medial prefrontal cortex) (Abubshait et al., 2023).

Even though this line of research did not address the question of adoption of the Intentional Stance towards the robot directly, it sheds light on robot behaviours (e.g. engagement in eye contact) that influence social attunement, which, in turn, might contribute to the adoption of the Intentional Stance—a topic to be explored in future research.

In a different, but related, line of research, Willemse et al. (2018) engaged participants in a joint attention task with a robot avatar on the screen. In this case, the study was a gaze-leading, rather than gaze-cueing, paradigm. This meant that the human observer was leading (rather than following) the gaze of the robot. In their study, Willemse and colleagues measured eye movement patterns of participants engaged in the task. Participants were asked to choose (at will) between two objects presented to them simultaneously. The iCub had two different 'identities'—Jimmy and Dylan. While Jimmy followed participants' gaze in 80% of trials (by directing its own gaze towards the same object as the one chosen by the participant), Dylan followed participants only in 20% of trials. The results showed that participants engaged in joint attention with the 'following' robot to a higher degree, but they also rated its human-likeness higher. This suggests that mutual reciprocity in social signals, such as gaze, are an important factor not only for social attunement but also for anthropomorphism.

5.3 Belief in Experience Sharing, Without Shared Goals

The next classification of interactive protocols encompasses experiments where humans believe that they share experiences with the robot. A belief in a shared experience (without shared goals) can be induced by engaging participants in a shared activity with a robot, like in Marchesi et al. (2022), or by sharing task space, like in Belkaid, Kompatsiari, et al. (2021) (cf. Fig. 5.1).

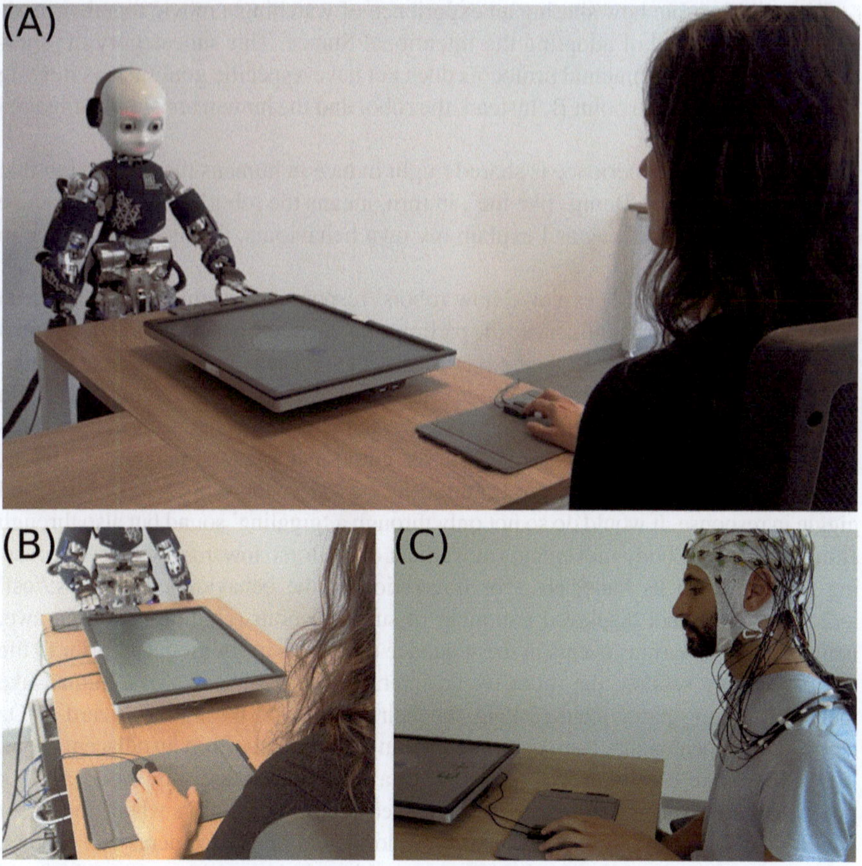

Fig. 5.1 An example of an interactive protocol involving a shared task space. (**a**) Participants played a strategic game called the *Chicken game* with the robot iCub (Belkaid et al., 2021). The players' individual payoffs depended on both their decisions, that is, whether each of them chose to stay on, or deviate from, the collision course. (**b**) The game was displayed on a screen placed horizontally between the participant and the robot. Both players acted on this shared environment by pressing the button corresponding to their choices. (**c**) This setup allowed collecting behavioural and neural data measured by EEG. The results showed that when iCub frequently avoided eye contact with them, participants were more sensitive to their outcomes—as indicated by outcome-related brain activity measured by EEG—and tended to make more choices based solely on their own payoffs without taking the other player's actions and payoffs into account—as signalled by the presence of more self-oriented patterns in their choice sequences. This experiment, even when not addressing the adoption of the Intentional Stance per se, provided insights into the key role of the robot's communicative behaviour (such as gaze) in how humans engage in their interaction with the robot, and to what extent they experience it as social

It is important to note that sharing experiences would presuppose that the robot actually has the capacity of 'experiencing'. Here, we do not consider this as a necessary condition for modulation of the adoption of Intentional Stance. As we will demonstrate with the results of experiments presented below, the crucial aspect of 'sharing' experiences is the *belief* in sharing the experiences with the robot. The empirical work presented here focuses on the impact on the human observer of certain robot behaviours (in an interaction) 'as if' the robot experienced the world with the human, without necessitating that the robot actually experiences the world.

We will illustrate how sharing an experience of watching a movie together modulates the likelihood of adopting the Intentional Stance. This subcategory of 'sharing experience' experimental protocols does not have a specific goal such as moving a table from point A to point B. Instead, the robot and the human are simply engaged in an activity together.

A belief that an experience is shared might induce in humans the impression that the robot is 'like me'. Being 'like me', in turn, means the robot's behaviours can be explained in the same way as I explain my own behaviours, that is, by reference to mental states.

Marchesi et al. (2022) explored how robots' responses to events, shared between the robot and the human, influenced the participants' inclination to adopt the Intentional Stance towards the robot. In their experiments, the researchers sat participants adjacent to the iCub robot and made participants watch a series of videos 'together' with the robot. In one condition, the robot was programmed to react to events in the videos in a way that its responses resonated with the emotional events of the movie. That is, for example, when a funny event occurred in the video, the robot would laugh and giggle in response. It would do so not only through a 'giggling' sound but also through entire human-like body movement (movement of the torso towards the screen, covering of its face in its hand, etc.; for illustration of the behaviours, see https://osf.io/2ckxv). The robot displayed a number of such reactions (giggling, being in awe, being scared) to various events in the sequence of videos. In addition, and prior to the video-watching session, the robot would interact with participants in a human-like manner (greeting participants, asking for their name, etc.) through a Wizard-of-Oz technique (its behaviours were controlled by an experimenter from an adjacent room. This human-like condition was presented to a group of participants, while another group of participants were exposed to a 'mechanistic' condition where the robot would display only purely mechanical behaviours (calibrating cameras and motors, etc.). It would not interact with participants, and in reaction to video events, it would only produce mechanical reactions, such as a beeping sound of a sensor.

Participants were asked to take part in the InStance Test (Marchesi et al., 2019) prior, and subsequent to, the interaction with the robot. The results showed that the score in the InStance Test increased after the interaction with the robot behaving in a human-like manner, but not when it showed mechanical behaviours. This effect was replicated several times, and thus, appears to be quite stable and robust (Marchesi et al., 2022, under review; Navare et al., 2022). The effect suggests that a short interaction with a humanoid robot, in which participants believe in sharing experience, and in which they might perceive the robot as 'like-me', owing to emotional reactions that resonate with participants' own reaction, increases the likelihood of adopting the Intentional Stance towards the robot.

5.4 Belief in Shared Experiences, with Shared Goals

The most 'interactive' or 'engaging' shared experience situation (thus, the 'highest' level of interactivity, in our taxonomy) is when the human not only believes that they share the experience with the robot, but they also actually share the goals of their task. Sharing goals means higher engagement of both partners, and perhaps more social attunement, as they need to coordinate and adapt to each other. This, in turn, might increase the likelihood of adoption of the Intentional Stance, because sharing a goal means that the partners are a 'team'. The assumption of being in the same team entails in-group attitudes, and the impression that in-group members are 'like me'. This paragraph will provide several examples of studies where a human was engaged with a robot in a task where they shared a goal. A collection of results from these studies demonstrates that sharing a goal with a humanoid robot increases the likelihood of adopting the Intentional Stance towards it.

One example of the case of sharing experience with outcomes and goals comes from Navare and colleagues (Navare et al., 2022). The authors engaged participants in a task in which they performed a complementary task with the robot. Importantly, the goal of the task could be achieved only jointly, while each partner was responsible for one part of the task. More specifically, the team was tasked to bring a cursor to a target location on the screen. One of the partners was responsible for moving the cursor to that location (by means of a sequence of key presses) while the other partner was tasked to confirm (with a key press) the target location once the cursor landed there. The likelihood was adopting the Intentional Stance was measured using InStance Test before, and after, the joint action task (similarly to the study of Marchesi et al., 2022). Results indicated that engaging in such shared task with iCub significantly increased participants' propensity to adopt the Intentional Stance towards iCub. This suggests that task sharing and joint goals promoted the perception of iCub's intentionality. Interestingly, in a follow-up study (Navare et al., 2024), the authors showed that increasing the likelihood of adopting the Intentional Stance towards iCub results in experiencing the sense of joint agency with the iCub—a phenomenon that has been observed in human–human interactions. This shows that the relationship between the Intentional Stance and social attunement might be bi-directional, and needs to be examined further to design robots well attuned to the human users in various contexts of social interactions.

The importance of perceiving the robot as a 'team member' has been further supported by the study of Abubshait and colleagues (Abubshait et al., 2021, under review). In their study, participants were completing a task, together with the iCub robot, in which they needed to classify objects based on colours into different bins. After doing so, they (and the robot) would receive points reflecting whether, or not, they classified the objects correctly. The key manipulation was that participants were either told that they were completing the task jointly and together with the robot (receiving points jointly with the robot), or they were completing the task separately (scoring individually). The results showed that participants in the 'collaborative' framing condition were more likely to adopt the Intentional Stance towards the robot as measured via the InStance Test. This effect persisted even after controlling for participants' baseline intentionality scores. This suggests that creating a context of 'teamwork' with the robot increases the likelihood of adopting the Intentional Stance, which might be due to the phenomenon of perceiving team members as 'in-group' members, and thus more 'like-me'. In other words, experiments and scenarios in which humans believe that they share the goals together with a robotic agent can induce an in-group bias effect, increasing the tendency to anthropomorphize and attribute intentional states to it. This is due to the fact that in-group members are often perceived as more human-like, and thus, more likely to have mental states (Waytz et al., 2010).

The body of work on joint actions with the robot highlights the importance of goal-oriented tasks in Human–Robot Interaction research. Tasks that involve cooperation could enhance a human's motivation and attention, making humans more receptive to robots' actions and allowing them to interpret robots as more human-like. This suggests that the nature of the task can significantly impact how humans perceive and interact with robots. As for applied implications for Human–Robot Interaction, this work suggests that designers of social robots could leverage framing effects (e.g., by manipulating interaction context) to influence how people perceive robots. By structuring interactions and tasks in specific ways, where humans and robots are to share tasks and goals together, designers can potentially encourage users to adopt the Intentional Stance, which might lead to more positive and engaging interactions with robots.

5.5 Conclusion

This chapter presented various degrees of 'interactivity' of experimental protocols that have been used to study how *interaction* per se affects the adoption of the Intentional Stance. Experimental protocols that operationalized the concept of sharing environmental stimuli through, for example, joint attention showed that there might be a dissociation between engagement in joint attention with the robot and adoption of the Intentional Stance—while the former takes time to develop, the latter diminishes as a function of repetitive and constant behaviour of the robot. Interestingly, the dissociation observed here is akin to the dissociation between implicit markers of attunement and explicit judgements of human-likeness reported in the studies that have been conducted as 'pure observation' (Chapter 4). Finally, experiments that involved the highest degree of 'interactivity' where participants believed in sharing experience with the robot show that engagement in shared activity—with or without a shared goal—and perception of a robot as a 'teammate' increase the likelihood of adopting the Intentional Stance. This collection of results not only deepens our understanding of Human–Robot Interactions and factors contributing to the Intentional Stance, but also lays a robust foundation for shaping the future landscape of human–robot collaborations, ensuring that these interactions are not only technologically sophisticated, but also elicit social attunement.

Interview with Bill Vorn

by Lorenzo Parenti and Ziggy O'Reilly

The presence first, it's very important. There is a physical presence that you don't have with virtual objects or creatures or entities. And it's a little bit like, when you listen to music, you can listen to the CD or you can go to the concert to see the show. And people like to go to a concert because they see the humans behind the music. With robots, it's the same thing. It's the real things, like going to theater. Why do we go to theater when we have cinema? But we still go. There's a physicality, there's an immediacy, and there's a presence that is not there with film.

– Bill Vorn

Based in Montreal, Bill Vorn is working in the field of Robotic Art for more than 30 years. His installation and performance projects involve robotics and motion control, sound, lighting, video, and cybernetic processes. He pursues research and creation on Artificial Life and Agent Technologies through artistic work based on the 'Aesthetics of Artificial Behaviors'. He holds a Ph.D. degree in Communication Studies from UQAM (Montreal) for his thesis on 'Artificial Life as Media'. He teaches Electronic Arts in the Department of Studio Arts at Concordia University (Intermedia program) where he is Full Professor.

His work has been presented in many international events, including Ars Electronica, ISEA, DEAF, Sonar, Art Futura, EMAF, and Artec. He has been awarded a Numix prize (2016, Montreal), the Vida 2.0 award (1999, Madrid), the Leprecon Award for Interactivity (1998, New York), the Prix Ars Electronica Distinction award (1996, Linz), and the International Digital Media Award (1996, Toronto). He has worked in collaboration with many Canadian artists (including

Edouard Lock, Robert Lepage, Gilles Maheu, Istvan Kantor, and Louis-Philippe Demers). He was cofounder of the electronic pop music band Rational Youth with Tracy Howe in 1981.

Summary
On the 16th of January 2024, two Ph.D. students, Lorenzo Parenti and Ziggy O'Reilly, connected to Zoom from Italy to speak with a robotic artist, Bill Vorn, in Montreal, Canada. The interview covers Bill Vorn's robotic art installations and performances. Key topics include anthropomorphism/biomorphism, the role of intuition in evoking emotional responses from the viewers, the role of movement and sound to create evocative performances, industrial aesthetics, accidental discoveries, physical embodiment, and interactivity.

Keywords
Robotic art, Robotic performance, Robot installations, Anthropomorphism, Physical presence, Appearance, Evocative robot performance

Conversation on the Interaction Between Robot Performers and Human Audience

Lorenzo Parenti In your work, you refer a lot to the concept of anthropomorphism. Could you describe this concept and the relevance that this concept has to your artistic practice?

Bill Vorn Sometimes I say anthropomorphism, but maybe it's probably better to say biomorphism. Of course, it depends on the robots that you make – if they look more like humans or more like animals or more abstract.

I would say, for me, it's more about the concept of projection, like what people are sort of transferring onto those objects, which is sort of the motivation behind the work. But there's a lot of things that are done intuitively. I cannot say that there's a very defined theoretical path to all this.

It's more you do it and about how you feel when you do it.
It's a bit like writing a song.
You just feel it, and you write it according to what you feel.
It's a bit the same thing with the robots.

I don't know in advance how people are going to feel in front of those entities. But I assume that what I feel is kind of close to what they perceive.

Lorenzo Parenti Okay, thanks. About this, do you think there are some viewers or observers that are more prone to anthropomorphize than others or let's say, perceive this biomorphism?

Bill Vorn Yeah, I think it's sort of a quality, but something that people experience at different levels, and not everybody sort of reacts the same way or has the same impressions. It's very different from one person to another. Sometimes you're going to have people who are very, very, sensitive and they know they're very prone to projecting things on the objects. And sometimes it's the opposite.

Lorenzo Parenti Thanks. I also wanted to ask you something else about the concept of simulacra. We found 'simulacra,' in your descriptions of your work. For instance, in the *Mega Hysterical Machine* (2010) and in the *Intensive Care Unit* (2021) work. Could you expand upon this concept of simulacra?

Bill Vorn Simulacra is in a sense a construct that is made to create some kind of a make-believe.

> ***Because of course we all understand that, that these are objects, machines, these are artificial things. But they play a game.***

They're part of a show or an installation or a performance in front of the public and they are presented in a context that makes them look like something. And also according to the theme of the work, because it's always a little bit different, but in the case of *Intensive Care Unit* (2021, cf. Fig. 5.2), if I'm talking about sick machines or a clinic of hospitalised machines, bedridden machines, and things like that, automatically people, have an idea, vague idea, of what it could be like, and by adding some props like the ECG displays or whatever, it finally looks a little bit like, you're in this machine hospital.

So that's the simulacra in this context. But it's not a real hospital or machine. It's a totally abstract place, a black box with the show going on. And also, sometimes other things that are not related at all to the concept of, in this case, a clinic of machines.

Lorenzo Parenti And do you think the concept of adopting either the Intentional or Design Stance applies also to the viewers in your performances? For instance, do you think that some of your performances could trigger the viewers to consider the machine as an intentional agent?

Bill Vorn I guess it could, but I'm not sure. It's because I'm not really looking for that. I'm not questioning anybody about that. And I would say that, in general, I'm more interested into the feeling that people get out of it.

> ***For me, the success of a piece is really about the internal impressions that the viewers get from being there at that moment in front of those machines doing their thing.***

They could be interactive or not, or it could be like a giant automaton. It doesn't matter. What's important is what they feel inside. And this could be a wide range of different feelings. Like it could be fear, joy, sadness, whatever. It just depends what it evokes for them.

Lorenzo Parenti Yeah, and it also comes to my mind the *DSM-VI* (2012) artwork. And in that case, you stated that the machines are abstract structures that are made of materials such as aluminum, plastic, and so on. But at the same time, you stated that despite being machines, it's their behaviors that give them this living aspect such as multiple personalities. So, it seems that from these artworks and from its description, that machines are somewhere in the middle between machine and living entities. Because on one hand, we put the focus on the fact that they're made of materials that are non-organic. On the other way, we're putting the focus on the fact that they're triggering also a perception of personalities. So, do you think that machines are always going to be in this middle ground?

Fig. 5.2 Intensive Care Unit (2021; top). Mega Hysterical Machine (2010; bottom left). DSM-VI (2012; bottom right)

Bill Vorn I would say there is sort of a mechanism that works very well, which is how people tend to look at the moving object and automatically start projecting sort of a soul into this object or some kind of autonomy or, a self -awareness of the object, just because it's moving.

And it will always reminds me of a project by one of my students a long time ago where he just put a little robot together with chicken bones from his last supper [laughs] or whatever, and a few motors, and he put that on the table.
And it was incredible.
From a technical point of view, it was very basic stuff.

If you look at the object made of bones and moving by itself we really had the impression it was alive, it was a creature of some kind, and at some point, it became a bit self-aware, we were a bit afraid of that.

Not everybody, but some people. [laughs].

This phenomenon is really kind of something I work with all the time. It's this reflex that people have towards moving objects and any kind of moving objects. Especially if they're not easily foreseeable or you can't really predict what they're going to do next.

Lorenzo Parenti Yeah, and actually I was about to ask you something related to this; basically, I think that maybe the way the object moves could be one of the most triggering features of the machine. And I was wondering if there is any other feature like that, that you saw?

Bill Vorn The visual aspect, of course, is very important because usually that's our first contact with the entity, even if it's not moving. So, we already have sort of an impression just by looking at it. But there's also all the, I would say, human-driven intentions that we put in our work, our machines, because we sometimes give them a name, we give them context. So, this is also part of the viewers experience.

Lorenzo Parenti Thank you. And on my side, I have one last question. I was very curious to know what's the inspiration or the process behind developing such a strong, and very triggering aesthetic?

Bill Vorn You mean in general or different pieces?

Lorenzo Parenti I'm actually mostly thinking about the *Intensive Care Unit*, because that's something that really stroke me visually. But in general, among all the machines and environments that you created; did it come from imagination or inspiration from a movie?

Bill Vorn It would be difficult to say that I've not been influenced by science fiction or movies alike and literature and comics and all this, like it all plays a part on where I'm coming from.

Of course, I was a musician before, making electronic music in the 80 s. So that's probably what influences me the most. But in terms of the robots, I would say, this is a mixture of different things. For example, the way I work, because I'm not a sculptor, I come from communication studies and music, so, no real background in visual arts, I've never done any kind of freehand sculpture stuff. I use industrial parts, the material I use, just comes out from the warehouse. And I don't really sculpt the materials, it's raw aluminum bars and things like that. That shapes the aesthetics a little bit of the robots. And I like it because it makes the machines look more like machines.

I'm not trying to transform the machine into something else.

In fact, it's perfect for me that you see it as a machine, as you consider it as a machine. Because then afterwards, you will start questioning yourself, like, "how come, I am projecting things onto this object?"

This bunch of, or this pile of, metal moving, suddenly starts to make you feel something.

That's really weird.

And that's pretty much what is behind the way I work and the way I kind of actualise my intentions.

Lorenzo Parenti Thank you very much. It's really super interesting. I have a bunch of other questions, but I will leave the floor to Ziggy.

Ziggy O'Reilly Thanks Lorenzo! So, I wanted to ask, what do you think the role is of physical embodiment on the performance, as opposed to perhaps using two-dimensional artificial agents?

Bill Vorn Well, I think it's quite important that there's a physicality to all this. Otherwise, I wouldn't make the robots [laughs]. It would save me a lot of problems.

But the presence first, it's very important. There is a physical presence that you don't have with virtual objects or creatures or entities.

And it's a little bit like, when you listen to music, you can listen to the CD or you can go to the concert to see the show. And people like to go to a concert because they see the humans behind the music.

With robots, it's the same thing. It's the real things, like going to theater. Why do we go to theater when we have cinema? But we still go. There's a physicality, there's an immediacy, and there's a presence that is not there with film.

And also, there are the other things, for example, like me, I never put any distance or fence or separation between the viewer and the machine. So that also creates a different relationship with the robot because somehow you could get hurt by the machine or you can touch it if you want.

Everything is allowed somehow. You're not in a zoo where the animals are behind bars. You're in their environment. You're disturbing them somehow. And you don't know how they will react.

Ziggy O'Reilly And do you think that it might change how emotionally the viewers respond to the robots, whether the robots are reciprocally interacting with each other?

Bill Vorn Well, it's a complex question. For me, the robots, they theoretically don't influence each other. They're all separate entities. But it could happen that there are some interactions.

For example, in the *DSM-VI* (2012) work, there is a robot and there is a face tracking algorithm. But sometimes it sees faces where there is no face at all. Because of the limitations of the algorithms, we could look at another machine, and then see a face and then start behaving as an autistic machine. And then, you can see the machine suddenly like, stopping, moving, looking at this other robot for a while and then doing something else or totally trying to avoid the other machine.

Ziggy O'Reilly Yeah, makes a lot of sense. And just now you said that some of these behaviors are accidental. Have you come across an accidental behavior, which led to an improvement in the performance in any of your works?

Bill Vorn Yeah, very often.

Ziggy O'Reilly What was your favorite accident?

Bill Vorn Oh, favorite, there's many. But for example, in *La Cour des Miracles* (1997) (Fig. 5.3) which is an old piece I did with Louis-Philippe Demers, there was a machine called the limping machine, because it was supposed to walk and limp at the same time. But it was not able to walk really, for some mechanical reason. And we just flipped it on its back and then it became a helpless robot trying to get back on its feet. And it was a lot more convincing like that than when it was trying to walk with difficulty. But with the same machine, in a different position, it created a different impression totally. The sub-theme of this work was the misery of the machine. So, it was obviously about making them look not so healthy.

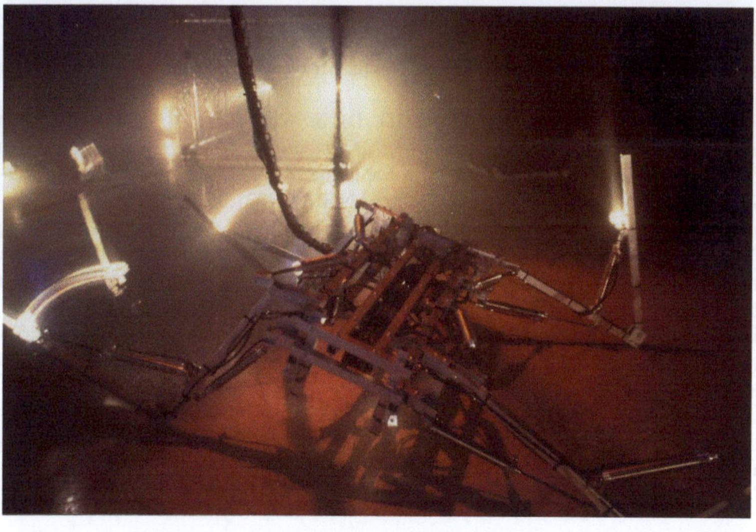

Fig. 5.3 La Cour des Miracles (1997), 'The Limping Machine'

Ziggy O'Reilly That must have been quite a nice thing to experience. Okay, I think for me, I've asked most of the really pressing questions that I had. I was just wondering if Lorenzo, there was anything else that you wanted to ask?

Lorenzo Parenti Just one final question. It's like the usual question that Ziggy and I do at the end of the interviews. And that is, if you have a favorite robot or in general a robot that you see as a social robot, a companion, in the future?

Bill Vorn I don't like robots enough to think about them like that, like a companion.

In terms of robots or show machines, for me it's always going to be Survival Research Labs performances, which are not really robots, they are all remote-control machines, which is somehow not so far from robots, but because they were kind of the first ones (well, if we forget about Jean Tinguely's, *Homage to New York in the 60 s,*[1] which was this big machines that sort of self-destruct, in front of the viewers). Survival Research Labs did a bit the same, but with a larger scale approach, and it was more a show, very noisy, very loud, and they started at the end of the 70 s, so it's a long time ago.

It really had a strong effect on me to see that. It really got me into working with machines and robots, even though I never really liked to work with mechanical stuff before.

I'm not the guy who repairs his car, or even his bicycle, like, I don't like to do that.

Lorenzo Parenti Thanks. Thank you very much. And thank you for your time.

References

Abubshait, A., & Wykowska, A. (2020). Repetitive robot behavior impacts perception of intentionality and gaze-related attentional orienting. *Frontiers in Robotics and AI*, 7. https://doi.org/10.3389/frobt.2020.565825

Abubshait, A., Perez-Osorio, J., Tommaso, D. D., & Wykowska, A. (2021). *Collaboratively framed interactions increase the adoption of Intentional Stance towards robots*. In 2021 30th IEEE international conference on robot human interactive communication (RO-MAN) (pp. 886–891). https://doi.org/10.1109/RO-MAN50785.2021.9515515.

Abubshait, A., Kompatsiari, K., Cardellicchio, P., Vescovo, E., De Tommaso, D., Fadiga, L., D'Ausilio, A., & Wykowska, A. (2023). Modulatory effects of communicative gaze on attentional orienting are driven by dorsomedial prefrontal cortex but not right temporoparietal junction. *Journal of Cognitive Neuroscience, 35*(10), 1670–1680.

Abubshait, A., Perez-Osorio, J., Tommaso, D. D., & Wykowska, A. (under review). Temporal interplay between cognitive conflict and attentional markers in social collaboration. *Psychophysiology.*

Bartneck, C., Kulić, D., Croft, E., & Zoghbi, S. (2009). Measurement instruments for the anthropomorphism, animacy, likeability, perceived intelligence, and perceived safety of robots. *International Journal of Social Robotics, 1*, 71–81.

[1] See https://vimeo.com/218619751

Belkaid, M., Kompatsiari, K., De Tommaso, D., Zablith, I., & Wykowska, A. (2021). Mutual gaze with a robot affects human neural activity and delays decision-making processes. *Science Robotics, 6*, eabc5044.

Friesen, C. K., & Kingstone, A. (1998). The eyes have it! Reflexive orienting is triggered by non-predictive gaze. *Psychonomic Bulletin & Review, 5*(3), 490–495.

Kompatsiari, K., Tikhanoff, V., Ciardo, F., Metta, G., & Wykowska, A. (2017). The importance of mutual gaze in human-robot interaction. *2*, 443–452. https://doi.org/10.1007/978-3-319-70022-9_44.

Kompatsiari, K., Ciardo, F., Tikhanoff, V., Metta, G., & Wykowska, A. (2018). On the role of eye contact in gaze cueing. *Scientific Reports, 8*(1), 17842. https://doi.org/10.1038/s41598-018-36136-2

Kompatsiari, K., Ciardo, F., De Tommaso, D., & Wykowska, A. (2019a). *Measuring engagement elicited by eye contact in human-robot interaction*. In 2019 IEEE/RSJ international conference on intelligent robots and systems (IROS) (pp. 6979–6985). https://doi.org/10.1109/IROS40897.2019.8967747.

Kompatsiari, K., Ciardo, F., Tikhanoff, V., Metta, G., & Wykowska, A. (2019b). It's in the eyes: The engaging role of eye contact in HRI. *International Journal of Social Robotics*. https://doi.org/10.1007/s12369-019-00565-4

Kompatsiari, K., Bossi, F., & Wykowska, A. (2021). Eye contact during joint attention with a humanoid robot modulates oscillatory brain activity. *Social Cognitive and Affective Neuroscience, 16*(4), 10. https://doi.org/10.1093/scan/nsab001

Kompatsiari, K., Ciardo, F., & Wykowska, A. (2022). To follow or not to follow your gaze: The interplay between strategic control and the eye contact effect on gaze-induced attention orienting. *Journal of Experimental Psychology: General*. https://doi.org/10.1037/xge0001074

Marchesi, S., Ghiglino, D., Ciardo, F., Perez-Osorio, J., Baykara, E., & Wykowska, A. (2019). Do we adopt the intentional stance toward humanoid robots? *Frontiers in Psychology, 10*, 450. https://doi.org/10.3389/fpsyg.2019.00450

Marchesi, S., Tommaso, D. D., Perez-Osorio, J., & Wykowska, A. (2022). Belief in sharing the same phenomenological experience increases the likelihood of adopting the intentional stance toward a humanoid robot. *Technology, Mind, and Behavior, 3*(3). https://doi.org/10.1037/tmb0000072

Marchesi, S., De Tommaso, D., Kompatsiari, K., Wu Y., & Wykowska, A. (under review). Tools and methods to study and replicate experiments addressing human social cognition in interactive scenarios.

Navare, U. P., Kompatsiari, K., Ciardo, F., & Wykowska, A. (2022). *Task sharing with the humanoid robot iCub increases the likelihood of adopting the Intentional Stance*. In 2022 31st IEEE International conference on robot and human interactive communication (RO-MAN) (pp. 135–140). https://doi.org/10.1109/RO-MAN53752.2022.9900746.

Navare, U. P., Ciardo, F., Kompatsiari, K., De Tommaso, D., & Wykowska, A. (2024). When performing actions with robots, attribution of intentionality affects the sense of joint agency. *Science Robotics, 9*, eadj3665.

Posner, M. I. (1980). Orienting of attention. *Quarterly Journal of Experimental Psychology, 32*, 3–25. https://doi.org/10.1093/acprof

Waytz, A., Gray, K., Epley, N., & Wegner, D. M. (2010). Causes and consequences of mind perception. *Trends in Cognitive Sciences, 14*(8), 383–388. https://doi.org/10.1016/j.tics.2010.05.006

Willemse, C., Marchesi, S., & Wykowska, A. (2018). Robot faces that follow gaze facilitate attentional engagement and increase their likeability. *Frontiers in Psychology, 9*, 70. https://doi.org/10.3389/fpsyg.2018.00070

Chapter 6
Factors Influencing the Adoption of the Intentional Stance: Culture

Serena Marchesi and Nicolas Spatola

> *Yolandi Visser [Reading]: All the sheep in Magooville were white,*
> *but Abel was a black sheep...[pauses].*
> *Yolandi Visser: Do you know what's a black sheep?*
> *Chappie: No.*
> *Yolandi Visser: It's like, when you're different from everyone else.*
> *Chappie: Like me. I look different from you and Amerika.*
> *Yolandi Visser: Yeah.*
> *Chappie: And the boys.*
> *Yolandi Visser: Yeah, but it's not so much how you look, it's special, like what's inside. That's what makes you different. See, it's who you really are... inside, your soul.*
>
> *– Blomkramp, N., CHAPPiE (2015), Columbia Pictures.*

6.1 Introduction

Besides individual differences discussed in Chapter 3, culture is another factor that falls within the umbrella of possible modulators of the Intentional Stance towards robots (Broadbent, 2017; Perez-Osorio & Wykowska, 2020). Cultural values and norms are deeply engrained in individuals throughout their development and upbringing, leading to the use of such values as biases and heuristics in society and daily life. This, in turn, shapes interactions with others, including artificial agents, such as robots.

S. Marchesi (✉)
Italian Institute of Technology, Social Cognition in Human-Robot Interaction Unit, Genova, Italy
e-mail: serena.marchesi@iit.it

N. Spatola
Artimon Perspectives, Paris, France

A. Wykowska (ed.), *Intentional Stance Towards Humanoid Robots*,
https://doi.org/10.1007/978-3-031-65483-1_6

For example, differences between Eastern and Western values, such as collectivism and individualism (Varnum et al., 2010; Triandis, 1993; Triandis & Gelfand, 2012), affect the way people approach others—while collectivistic cultures tend to prioritize harmony and positive social interactions, individualistic cultures emphasize self-reliance (Hofstede, 2011). This affects not only interactions among humans, but also various aspects of interactions with robots, as people with different cultural backgrounds might vary in their behaviours towards robots, thus exhibiting different levels of likeability, satisfaction, trust, engagement, and interpersonal distance during interactions with them (Lim et al., 2021).

Moreover, cultural differences influence people's tendency to anthropomorphize robots and attribute intentionality to them (Spatola et al., 2022). For instance, some cultures may be more prone to ascribe anthropomorphic traits to robots. Factors such as exposure to robots, media exposure, and technological development can modulate this tendency (Airenti, 2018; Bartneck et al., 2006). Also, historical and religious backgrounds may be concurrent reasons for cultural differences in anthropomorphism (Shaman et al., 2018). As an example, Japanese culture's association with high anthropomorphism can be attributed to the beliefs in animism prevalent in the dominant Shinto region (MacDorman et al., 2009).

Understanding, and appreciating, these cultural differences is fundamental to design robots which positively resonate with individuals from diverse cultural backgrounds. Cultural sensitivity is indispensable in ensuring that social robots successfully navigate diverse social contexts and foster meaningful connections with their human counterparts, thus promoting more inclusive and positive interactions that match individuals' unique expectations and needs (Papadopoulos et al., 2022; Bruno et al., 2019). Conversely, neglecting the impact of cultural values and norms during the design of a social robot can lead to mistakes and misunderstandings, ultimately resulting in negative experiences for the human users.

With this in mind, we conducted a series of studies to explore whether, and how, individuals' cultural values and norms affect their social attunement when engaged in an interaction with a robot. As discussed earlier in this book, social attunement is intrinsically related to the adoption of the Intentional Stance, as through attunement, we might perceive robots more 'like me', and thus also intentional agents (see Chap. 5, study by Marchesi et al., 2022 in which social attunement during movie watching and perception of the robot 'like me' increased the likelihood of the adoption of the Intentional Stance). In the first part of this chapter, will focus on two socio-cognitive mechanisms that can be subsumed under the large umbrella concept of social attunement, namely joint attention and social inclusion. We will discuss how those mechanisms are modulated by cultural background.

6.2 Joint Attention with a Robot Across Cultures

Marchesi and colleagues (Marchesi et al., 2023a) adapted a traditional screen-based gaze-cueing paradigm with a robot agent. The directional cue was delivered after a manipulation of gaze contact: the robot's gaze was either straight-ahead, towards the participants, or averted, prior to the directional gaze cue. Marchesi and colleagues conducted the same experiment across Italy and Singapore.

Results showed higher degree of engagement in joint attention with the robot in the Singaporean sample compared to the Italian sample, regardless of the gaze type (eye contact or no eye contact). In addition, participants' ratings of engagement with the robot agent after each block of the task provided further insights. While Italian participants rated the avatar more engaging during the blocks with eye contact, compared to those with no eye contact, Singaporean participants did not demonstrate any significant difference in engagement between the two types of gaze.

Overall, these findings prompted an intriguing discussion regarding cultural variations in joint attention with (robot) agents on the screen, as well as differences in engagement with eye contact. It seems that while participants from Singapore are likely to engage in joint attention with the robot agent, for the Italian participants a robot's face displayed on the screen might not be sufficient. This discrepancy might be explained by the fact that the perception of a simplified 2D robot agent as a social stimulus might not be the same across those two countries, thus appearing to be culturally dependent. Although the study did not directly assess the adoption of the Intentional Stance as a function of culture, engagement in joint attention is a crucial mechanism of social attunement that is highly related to the Intentional Stance. The fact that the Singapore sample showed engagement in joint attention with 2D representations of a robot, while the Italian sample showed this effect to a lesser extent (which consistently showed engagement in joint attention with a physically embodied robot, cf. Kompatsiari et al., 2018, 2022), it might suggest that individuals with Asian cultural background require less from the interaction partner (are more minimalistic) than individuals with Western background, to engage their own socio-cognitive mechanisms in an interaction.

Following up on the screen-based intercultural research related to socio-cognitive mechanisms in relation to robots, Marchesi et al. (2023b) set out to examine cultural differences in engagement in joint attention with a physically embodied iCub. Marchesi and colleagues implemented in Singapore a gaze-cueing study that consisted of an identical protocol to that of Kompatsiari and colleagues (Kompatsiari et al., 2022), conducted in Italy. The results between Singapore and Italy in the joint attention task were comparable. This was at odds with the results of the study with the robot agent on the screen, as described earlier, but it might actually be taken as supporting evidence for the claim that the Asian sample requires 'less' from the interaction partner to engage socially with it. That is, when the interaction partner is a 2D representation of a robot on the screen, this is not sufficient for the Western cultures to engage their socio-cognitive mechanisms of social engagement. On the contrary, individuals from Asian cultures might be willing to engage even with such 'minimalistic' social partner. When the robot becomes physically embodied—this is when also individuals from Western cultures are ready to engage in social interaction, and thus, their level of engagement in joint attention becomes comparable to that of individuals from Asian cultures. However, this speculative interpretation of the results needs to be further supported by future research using various experimental paradigms and across various cultures (extending the comparison between only Italy and Singapore).

In another study that targeted the adoption of the Intentional Stance directly, Marchesi and colleagues (Marchesi et al., 2023b) replicated the same study as reported in Chapter 5 of this book (Marchesi et al., 2022), namely the shared experience of movie watching with iCub. As the reader might remember from the previous chapter, the original study, which took place in Italy, comprised experiments in which participants watched videos together with iCub, which was programmed to respond either in a human-like manner or in a machine-like manner to events presented in the videos (Marchesi et al., 2022). The findings from this series of experiments revealed that, when iCub behaved in a human-like way, participants' likelihood of adopting the Intentional Stance increased. However, when iCub behaved as a machine, participants' tendency to adopt the Intentional Stance remained constant throughout their interaction with the robot.

When running the same experiment in Singapore, marchesi and colleagues replicated the same pattern of results, that is, a significant increase in participants' likelihood of adopting the Intentional Stance only when the robot behaved in a human-like manner (Marchesi et al., 2023b). This shows that social bonding and shared phenomenal experience, and possibly perception of the robot as being 'like-me', are powerful mechanisms that evoke the adoption of the Intentional Stance, independent of the cultural setting. Again, in line with the reasoning related to comparable results on joint attention across cultures, this set of studies demonstrates that physical embodied presence of a humanoid robot is presumably sufficient to engage socio-cognitive mechanisms also in the Western cultures.

6.3 Social Inclusion of Robots Across Different Cultures

Regarding social inclusion, past research showed that this phenomenon extends from human–human interaction (HHI) to HRI (Rosenthal-von der Pütten & Bock, 2023). In this context, our particular interest was to assess how cultural differences shape the way people include robots in their social sphere, namely their propensity to perceive robots as in-group social partners. We developed a modified version of the Cyberball game, a virtual ball-tossing game that has been extensively used in social psychology to investigate social inclusion (Williams & Jarvis, 2006). In the standard version of the Cyberball game, participants play with two other players depicted as animated icons on the screen. What varies in the game is the number of tosses received by participants, and thus the extent to which they feel to be (socially) included in the game. Indeed, ostracised participants receive fewer ball tosses compared to the other players. Conversely, included participants repeatedly receive the ball. Although the experimenters portray the task to participants as merely a way to train their mental visualisation skills, results show that participants typically care about the extent to which the other players include them in the game. Specifically, if participants are over-included (i.e. they receive the ball in half of the total tosses), or simply included (i.e. they receive the ball in one-third of the total tosses), they report more positive feelings—in terms of meaningful existence, perceived control, and self-esteem, compared to ostracised participants (i.e. the ones who receive the ball only in one-sixth of the total tosses) (Williams & Jarvis, 2006). Thus, the Cyberball game allows researchers to create a controlled environment in which it is possible to systematically investigate whether, and how, social inclusion is modulated by various factors, such as the number of tosses received by each player and/or the nature of the players involved in the game (Rosenthal-von der Pütten & Bock, 2023).

Marchesi and colleagues (Marchesi et al., 2021a) adapted the original Cyberball game to an HRI context and performed an online experiment in which participants played the game with another human and the iCub robot, both depicted as icons on the screen. The authors examined the impact of cultural differences, in terms of nationality and individual cultural stance, on people's tendency to socially include the robot. Two samples of participants varying in their nationality were recruited. On the one hand, Chinese participants were chosen as representative of a collectivistic, East Asian culture, whereas UK participants were chosen as representative of an individualistic, Western culture. Moreover, the authors examined whether the individual propensity to attribute intentionality to robots played a role in modulating participants' level of social inclusion exhibited towards iCub.

The findings revealed that individuals' cultural stance towards collectivism and their propensity to attribute a 'mind' to robots emerged as significant predictors of the degree of social inclusion demonstrated towards the iCub robot. In other words, the more participants displayed a collectivistic stance, and were disposed towards ascribing intentionality traits to the robot agent, the more they tended to include the robot in the Cyberball game (Marchesi et al., 2021a). This is an extremely interesting finding as it shows that acceptance of robots as in-group members might depend both on cultural stance and on the degree of adopting the Intentional Stance.

6.4 Cultural Differences, Intentional Stance, and Anthropomorphism

To deepen the answer to the question of relationship between the adoption of the Intentional Stance towards robots and individuals' cultural background, Marchesi and colleagues (Marchesi et al., 2021b) asked 600 participants from diverse countries to complete a series of tests, and their responses were analysed using path model analysis. This analysis allows for identifying the influence and directionality of all theoretical constructs involved in an experimental design. The findings revealed that collectivism had a significant direct impact on anthropomorphism, however, not on the adoption of the Intentional Stance. The authors also observed that anthropomorphism played a mediating role between collectivism and the adoption of the Intentional Stance, suggesting that, at the cross-cultural level, the varying degrees of anthropomorphic attributions influence people's likelihood of adopting the Intentional Stance towards humanoid robots.

Building on these results, Spatola and colleagues (Spatola et al., 2022) put forward a model (cf. Fig. 6.1) to describe the relationship between several concepts addressed in this book (cf. Chapter 2). Specifically, the Integrative Framework of Anthropomorphism (IFA) comprised three interconnected levels: the cultural aspects, the likelihood of adopting the Intentional Stance, and the individual inclinations to anthropomorphize (including ascriptions of intention, emotion, and cognition). In the model and in the empirical data, anthropomorphism was intrinsically defined as a complex concept that should be considered in a cultural, individual, and attributional context.

From a general viewpoint, the IFA highlights two types of processes. The first one is as a process of categorising a robot on the humanisation continuum. It determines if a robot is 'like a human'. The closer to the human, the higher the attribution of intentions, cognition, and emotions, as those are human characteristics. The second process, (partially) independent of the 'human-like' categorization, relies on the ascription of a mind in two correlated forms: mentalisation (closely related to the Intentional Stance) and spiritualism. Mentalisation is the manner of explaining behaviour. Spiritualism is the idea that a robot shares commonalities with other living beings populating the world. This subdivision echoes Fisher's view of anthropomorphism. Fisher proposed that anthropomorphism could be divided into an interpretative (i.e., situational explanation process) and an imaginative (i.e., general representation) form. Mentalizing (Intentional Stance) would be the interpretative form and spiritualism would be the imaginative form of mind attribution to non-human agents, such as robots.

The model (built on empirical data) demonstrates that the attribution of intention, emotion, and cognition to robots could depend on multiple tendencies in parallel, arguing that anthropomorphism is a complex, rather than a unitary, process—not only in terms of motivational factors, as posited by Epley and colleagues (Epley et al., 2007), but also in terms of processes underlying anthropomorphic representation of a non-human agent. For instance, attributing an intention to a robot may result from a social categorisation process (humanisation) or/and an interpretation

Fig. 6.1 Visual representation of the Integrative Framework of Anthropomorphism (IFA). (Spatola et al., 2022)

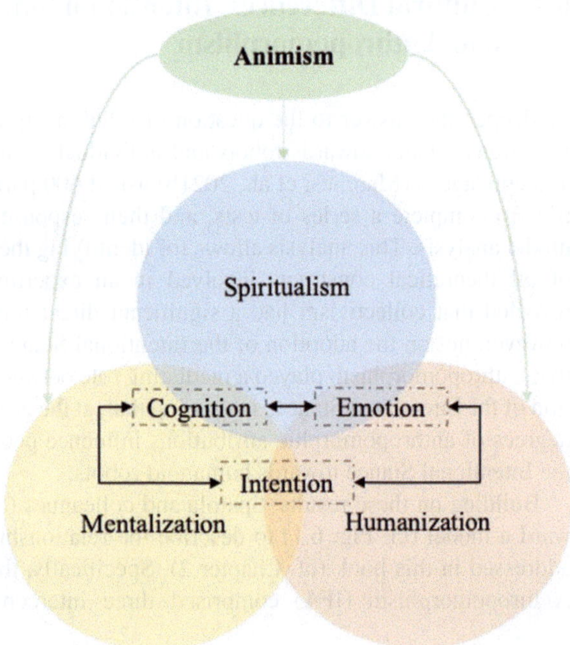

process (mentalisation). Therefore, in research on anthropomorphism, we should take these epistemological distinctions into account.

Interestingly, Spatola et al. (2022) found two different patterns when modelling the data from the Western and East Asian samples separately. For Western participants, animism was related to humanisation while for East Asian participants, animism was related to mentalisation. As we hypothesized, Western cultures proved to be more anthropocentric than East Asian cultures. The difference between cultures in how animism affects anthropomorphism—through either humanisation (the West) or mentalisation (the East)—illustrates that anthropomorphism might have different (culturally flavoured) facets. This confirms Urquiza-Haas and Kortschal's theory (Urquiza-Haas & Kotrschal, 2015) which highlights the interplay between cultural differences and individual variability as a crucial process in anthropomorphism. In terms of more methodological considerations, these results demonstrate that comparing anthropomorphic tendencies in the various questionnaires or tasks available in HRI literature may result in misleading interpretation due lack of delineation of constructs. The results of Spatola and colleagues show that concepts may be epistemologically different and attributing mental properties to a robot in an anthropocentric culture is not the same as in a culture with less anthropocentric values.

In other words, the question 'Does this robot have a mind?' in an anthropocentric culture would be closer to the question 'To what extent is this robot like a human?', while in a non-anthropocentric culture, the same question would be closer to 'Can this robot think and have emotions?'

6.5 Conclusions

This chapter described a collection of work that addressed cultural differences (and similarities) in aptitude for social attunement with robots, and towards adopting the Intentional Stance. Studies that have been conducted across labs in the Western part of the world (Italy) have been compared to analogous studies conducted in Singapore. The results suggest that the Asian cultures might require less from an interaction partner to engage their socio-cognitive mechanisms in an interaction. The Western cultures might be more likely to socially attune with physically embodied robots, and less so with 2D representations thereof. However, once an interaction involves a physically embodied robot, the socio-cognitive mechanisms evoked by the robot do not differ across cultures, which is also an important example of replicability of experimental design (and results) across labs and continents.

Finally, the last part of the chapter discussed conceptualisation of Intentional Stance and anthropomorphism that might differ across cultures, dependent on cultural values. Interestingly, more anthropocentric cultures (such as the West) might interpret questions about intentionality or mind in relation to a human, while in non-anthropocentric cultures this question might be interpreted very differently. In sum, it is important to highlight that cultural differences and adherence to various cultural values need to be taken into account not only in the design of robots but also in methodological approaches and the way one builds questionnaires and surveys.

Interview with Prof. Antonio Sgorbissa

by Lorenzo Parenti and Ziggy O'Reilly

We focus on the general attitude, and how to represent the fact that an individual person, although they share some attributes with the culture they identify with, they still have individual qualities. Then the model was developed to represent this kind of knowledge without falling into stereotypes, which is the biggest danger.

– Prof. Antonio Sgorbissa

Professor Antonio Sgorbissa is a computer scientist with a Ph.D. in Robotics. He is Full Professor at the Department of Computer Science, Bioengineering, Robotics, and Systems Engineering at the University of Genova. He served as the scientific coordinator of CARESSES—Culture Aware Robots and Environmental Sensor Systems for Elderly Support—from 2017 to 2020. CARESSES is an international project including European and Japanese labs that aimed to develop robots with cultural competence, capable of adapting their behaviour to the cultural identity of the user they are interacting with. Prof. Sgorbissa's research activities primarily focus on Robotics and Ambient Intelligence, with a special emphasis on methodological and technological aspects, as well as the cultural differences in social robotics.

Summary
On the 19th of February 2024 two Ph.D. students, Lorenzo Parenti and Ziggy O'Reilly, spoke with Prof. Sgorbissa. The interview covers anecdotes related to individual and cultural differences, attribution of mental states towards robots, and avoiding cultural stereotypes in programming robots.

Keywords
Culturally-competent robots, Cultural differences in social robotics

Conversation on the Role of Cultural Differences in the Design of Social Robots

Lorenzo Parenti So first, we would like to ask whether you think there are cultural differences in responding to a robot?

Antonio Sgorbissa Okay, first of all, let's start from the general idea of this project called CARESSES. It is about making culturally aware robots that are capable of adapting their behavior depending on the cultural identity of the person they are interacting with.

So this was the general idea.

Then of course, all the work was heavily influenced by the perspective of transcultural nursing and our perspective was more in this direction of cultural competence in healthcare. So, the behaviour of the robot (as the behaviour of nurses), should be culturally competent.

> ***How should a culturally competent robot behave? Of course, this culturally competent robot should consider the fact that there are different cultures.***

Meaning that if you're talking with an Italian person and you talk about football, this may be familiar to this person. If you talk with a person in India, they may be familiar with cricket. In Italy, you might think that not many people are familiar with cricket, right? But there still might be a person in Italy, who is familiar with cricket.

> ***We focus on the general attitude and how to represent the fact that the individual person, although they share some attributes with the culture they identify with, they still have individual qualities. Then the model was developed to represent this kind of knowledge without falling into stereotypes, which is the biggest danger.***

Lorenzo Parenti Yeah, it's super interesting and I guess that it's a huge problem, this kind of bias that affects participants, as well as researchers in the field, right? The follow-up question that I wanted to ask you is whether you can exploit biases and stereotypes in the robot design?

Antonio Sgorbissa Yeah, yeah. For example, at some point we were doing something with an English man, living in a care home, and the robot started asking him about life during World War 2.

Which, of course, as you can imagine, for English people is a particularly relevant topic.

He said later that he was very excited to have the possibility of telling his story of when he was a kid and his father was a prisoner in Iraq and it was very hard for him.

So, I think that there are cultural aspects that can really make the difference.

Also, during the CARESSES project, all of my English partners, without exception, referred to Pepper as "she," while all of my Italian colleagues referred to it as "he."

In Italy, this might be influenced by grammar: we use masculine attributes and articles with the word "robot," such as "un" (a) and "bel" (beautiful), and we translate "pepper" as "il pepe," again using a masculine article.

My English colleagues believed that Pepper is a "she" due to its "curvy shape," because "she takes care," and perhaps also because Pepper is considered a female name in English (consider, for example, Iron Man's former secretary, and then wife). These are just speculations, but to me, they present an intriguing anecdote.

Then of course, to confirm these cultural differences, we should make experiments with a larger sample of participants and so forth.

Ziggy O'Reilly Thank you for those interesting anecdotes! How are you deciding on the knowledge base that a robot might use if you're speaking to someone from the UK, for instance?

Antonio Sgorbissa So basically, the idea is that the knowledge base provides the information about the most probable topics.

The point is that the robot can use the knowledge base to make guesses and checks if the robot's assumption is true for this specific culture.

The idea is that you're in the interaction and the robot starts from some assumption. For example, the assumption that Italians are more familiar with having milk and cookies for breakfast than eggs and bacon. Then it is ready to revise this assumption.

> ***What is important is not only the knowledge base, but also the process that allows the robot to use this knowledge base in a non-stereotyped way, which is really the thing we want to avoid.***

Another aspect, which was part of it, is that the knowledge base is also coupled with a distribution of probability.

So, if the robot is talking to an Italian person and somehow discovers that this person is familiar with, or is interested in, Japanese culture, for example, they like Haiku poetry, then it increases the probabilities that other topics of Japanese culture may also be relevant for this person (Japanese anime movies, maybe?), and it will use this information.

You have this distribution of probability which tells you whether it is a good idea for the robot to talk about this thing with the person in this way. This distribution of probability changes dynamically depending on the specific person the robot is interacting with.

So yes, it's the core of the idea of how we represent culture.

I said "culture", but I should rather say "diversity" because recently we moved towards a more general concept of diversity, which includes culture as one of the dimensions.

And the idea is to use the same concept to talk with persons of different ages, various sexual orientations, various gender identities and various physical and mental capabilities.

Ziggy O'Reilly Okay, yeah, I understand. Thank you. It seems like your research is going in a very exciting direction.

So far, we've discussed that the robot needs to have a necessary knowledge base that is relevant and resonates with the specific individual that it's speaking to.

In addition to this knowledge base, do you think that non-verbal communication is important in the design of the robot?

Antonio Sgorbissa Given that our perceptions are strongly influenced by robots and other animated objects we encounter in movies (especially animated ones), I wonder about the expectations different cultures have for robot behaviour.

Do Indians look for "Indian-like" non-verbal behaviour, or do they look for "R2D2-like" and "Wall-E-like" behaviour in robots?

Consider that animated movies, like Pixar's *Wall-E* and Disney's *Beauty and the Beast*, feature many animated objects that adhere to strict rules of animation. Therefore, I speculate that when interacting with robots, people might look for the well-known stereotypes of robot behaviour, rather than seeking something that mimics human behaviour. Robot-like behaviour is well established, and we seem to seek it out!

Again, this is merely a hypothesis, and I don't strongly advocate for it. However, I believe it's a hypothesis worth exploring.

Lorenzo Parenti I wanted to ask you whether you have considered the attribution of mental states in your research or whether you have encountered this topic while doing your research? For instance, people thinking that the robot has good intentions, or it has a goal "in mind"?

Antonio Sgorbissa Okay. First, I have never done any experiments on this, so everything I will say is just my impressions.

So, my impressions are that everybody attributes these kinds of mental states to the robots, especially as soon as they are humanoid robots, but even this is not required.

I remember very clearly, when I participated in the RoboCup'99 competition in Stockholm (RoboCup is a soccer competition between robots, cf. Fig. 6.2). We had these robots which were not humanoids. The robots had no face. Nothing that can make them similar to humans or even animals.

But you can see during the competition that all researchers, were screaming to the robots something like "go go", or screaming to them with disappointment "why aren't you going there?!" I mean it was like they were screaming to living beings.

So, I'm pretty sure that there is no specific relation to being a humanoid, but maybe the behaviour, in that case, chasing the ball, was sufficient to make you think that the robot had mental states, such as intentions.

Fig. 6.2 Image of the Robocup 1999

There is another example that I always remember because it made me feel very nervous.

We were doing experiments with older people in care homes (Fig. 6.3). They had their robot (a Pepper robot) for 18 h. They were alone in the room for 6 sessions of 3 h each.

(This is very important because after a while probably they forgot they were part of an experiment).

And one of these participants, he was a man, the Englishman that talked about World War 2.

At some point, he wanted to talk about his wife. His dead wife.

And the robot… at the time, there was a bug, and it could not understand the keyword *wife*. We used keywords, because ChatGPT was not available yet.

But he insisted, because maybe he thought the robot was not understanding.

We were observing from outside.

I don't know how to say it in English, but our hearts were becoming smaller and smaller.

Can you imagine the situation? On the one hand, it was beautiful for the experiment and it was very moving also for this man.

Then we fixed the bug and he could talk about his wife.

Then, there was an Indian woman—when we took the robot away….

Of course, we had a detachment protocol because you cannot make experiments with older people in care and just say goodbye, right?

We had this protocol, but at the end of the day we took the robot away.

> ***And this woman, she really created a bond with this robot, which was incredible. She always said, "Pepper is my best friend."***

When the robot was taken away, she went and complained with the manager of the care home, because she was really upset that her friend, "Pepper," was taken away from her.

Fig. 6.3 Images from the CARESSES project

And as far as I know, this creates a very serious problem.

But the feeling we had was that she wasn't complaining, as she would be if a television set was taken away.

She was complaining as if a partner for conversation was taken away.

Lorenzo Parenti These kinds of anecdotes and stories are very insightful, thank you. We also wanted to ask you what's your favorite robot to work with?

Antonio Sgorbissa Okay, in my lab at the moment we have a teleoperation robot by Double Robotics, Husarion ROSbots, and plenty of quadcopters (Fig. 6.4).

We currently have a Navel robot by Navel Robotics. We have a Buddy robot. I have NAO, I have Pepper. I have an AlterEgo robot that is developed by the Italian Institute of Technology (IIT). I have a Spot robot—the quadruped robot. I think you give intentional states to Spot because it really looks like a dog and people talk about Spot as if it *wants* to go somewhere.

Fig. 6.4 Top: Robots and Intelligent systems of Citizens and the Environment lab (RICE) with drones, ROSbot (Husarion), Professor Einstein™ (Hanson Robotics), Buddy Robot (Blue Frog Robotics), Pepper (Softbank Robotics), Spot (Boston Dynamics), and NAO (Softbank Robotics)

So, it may be due to the fact that I worked for a longer time with it, but I really like the way Pepper is done. And I see that everybody immediately loves it. So, I always have to say to people, "okay, it's not Pepper. If you take Pepper out of the box, it's doesn't do anything. It's our system that makes it interesting." This is what I try to convince the people about, but it's true that people are very attracted by the smooth movements of Pepper. So, I think that I like it very much.

Lorenzo Parenti Yeah. That's super nice.

Ziggy O'Reilly Thank you very much for your time.

Antonio Sgorbissa You're welcome. My pleasure.

References

Airenti, G. (2018). The development of anthropomorphism in interaction: Intersubjectivity, imagination, and theory of mind. *Frontiers in Psychology, 9*, 2136. https://doi.org/10.3389/fpsyg.2018.02136

Bartneck, C., Suzuki, T., Kanda, T., & Nomura, T. (2006). The influence of people's culture and prior experiences with Aibo on their attitude towards robots. *AI & SOCIETY, 21*(1–2), 217–230. https://doi.org/10.1007/s00146-006-0052-7

Broadbent, E. (2017). Interactions with robots: The truths we reveal about ourselves. *Annual Review of Psychology, 68*, 627–652. https://doi.org/10.1146/annurev-psych-010416-043958

Bruno, B., et al. (2019). The CARESSES EU-Japan project: Making assistive robots culturally competent. In N. Casiddu, C. Porfirione, A. Monteriù, & F. Cavallo (Eds.), *Ambient assisted living. ForItAAL 2017* (Lecture notes in electrical engineering) (Vol. 540). Springer. https://doi.org/10.1007/978-3-030-04672-9_10

Epley, N., Waytz, A., & Cacioppo, J. T. (2007). On seeing human: A three-factor theory of anthropomorphism. *Psychological Review, 114*(4), 864–886. https://doi.org/10.1037/0033-295X.114.4.864

Hofstede, G. (2011). Dimensionalizing cultures: The Hofstede model in context. *Online Readings in Psychology and Culture, 2*(1). https://doi.org/10.9707/2307-0919.1014

Kompatsiari, K., Ciardo, F., Tikhanoff, V., Metta, G., & Wykowska, A. (2018). On the role of eye contact in gaze cueing. *Scientific Reports, 8*, 17842. https://doi.org/10.1038/s41598-018-36136-2

Kompatsiari, K., Ciardo, F., & Wykowska, A. (2022). To follow or not to follow your gaze: The interplay between strategic control and the eye contact effect on gaze-induced attention orienting. *Journal of Experimental Psychology: General, 151*(1), 121. https://doi.org/10.1037/xge0001074

Lim, V., Rooksby, M., & Cross, E. S. (2021). Social robots on a global stage: Establishing a role for culture during human–robot interaction. *International Journal of Social Robotics, 13*(6), 1307–1333. https://doi.org/10.1007/s12369-020-00710-4

MacDorman, K. F., Green, R. D., Ho, C. C., & Koch, C. T. (2009). Too real for comfort? Uncanny responses to computer generated faces. *Computers in Human Behavior, 25*(3), 695–710. https://doi.org/10.1016/j.chb.2008.12.026

Marchesi, S., Roselli, C., & Wykowska, A. (2021a). Cultural values, but not nationality, predict social inclusion of robots. In H. Li, S. S. Ge, Y. Wu, A. Wykowska, H. He, X. Liu, D. Li, & J. Perez-Osorio (Eds.), *Social robotics* (Vol. 13086, pp. 48–57). Springer. https://doi.org/10.1007/978-3-030-90525-5_5

Marchesi, S., Spatola, N., & Wykowska, A. (2021b). The mediating role of anthropomorphism in adopting the Intentional Stance towards humanoid robots. https://doi.org/10.31234/osf.io/zdr9v.

Marchesi, S., De Tommaso, D., Perez-Osorio, J., & Wykowska, A. (2022). Belief in sharing the same phenomenological experience increases the likelihood of adopting the Intentional Stance toward a humanoid robot.

Marchesi, S., Abubshait, A., Kompatsiari, K., Wu, Y., & Wykowska, A. (2023a). Cultural differences in joint attention and engagement in mutual gaze with a robot face. *Scientific Reports, 13*(1), 11689. https://doi.org/10.1038/s41598-023-38704-7

Marchesi, S., De Tommaso, D., Kompatsiari, K., Wu, Y., & Wykowska, A. (2023b). Tools and methods to study and replicate experiments addressing human social cognition in interactive scenarios. Preprint available at: https://doi.org/10.31234/osf.io/xyhwt.

Papadopoulos, C., Castro, N., Nigath, A., Davidson, R., Faulkes, N., Menicatti, R., et al. (2022). The CARESSES randomised controlled trial: Exploring the health-related impact of culturally competent artificial intelligence embedded into socially assistive robots and tested in older adult care homes. *International Journal of Social Robotics, 14*(1), 245–256. https://doi.org/10.1007/s12369-021-00781-x

Perez-Osorio, J., & Wykowska, A. (2020). Adopting the intentional stance toward natural and artificial agents. *Philosophical Psychology, 33*(3), 369–395. https://doi.org/10.1080/09515089.2019.1688778

Rosenthal-von der Pütten, A. M., & Bock, N. (2023). Seriously, what did one robot say to the other? Being left out from communication by robots causes feelings of social exclusion. *Human-Machine Communication, 6*(1), 7. https://doi.org/10.30658/hmc.6.7

Shaman, N. J., Saide, A. R., & Richert, R. A. (2018). Dimensional structure of and variation in anthropomorphic concepts of god. *Frontiers in Psychology, 9*, 1425. https://doi.org/10.3389/fpsyg.2018.01425

Spatola, N., Marchesi, S., & Wykowska, A. (2022). Different models of anthropomorphism across cultures and ontological limits in current frameworks the integrative framework of anthropomorphism. *Frontiers in Robotics and AI*, 16.

Triandis, H. C. (1993). Collectivism and individualism as cultural syndromes. *Cross-Cultural Research, 27*(3–4), 155–180. https://doi.org/10.1177/106939719302700301

Triandis, H. C., & Gelfand, M. J. (2012). A theory of individualism and collectivism. In P. A. M. Van Lange, A. W. Kruglanski, & E. T. Higgins (Eds.), *Handbook of theories of social psychology* (pp. 498–520). Sage.

Urquiza-Haas, E. G., & Kotrschal, K. (2015). The mind behind anthropomorphic thinking: Attribution of mental states to other species. *Animal Behaviour, 109*, 167–176. https://doi.org/10.1016/j.anbehav.2015.08.011

Varnum, M. E., Grossmann, I., Kitayama, S., & Nisbett, R. E. (2010). The origin of cultural differences in cognition: The social orientation hypothesis. *Current Directions in Psychological Science, 19*(1), 9–13. https://doi.org/10.1177/0963721409359301

Williams, K. D., & Jarvis, B. (2006). Cyberball: A program for use in research on interpersonal ostracism and acceptance. *Behavior Research Methods, 38*, 174–180. https://doi.org/10.3758/BF03192765

Chapter 7
On the Relationship Between the Intentional Stance and Robot Responsibility

Ziggy O'Reilly and Serena Marchesi

> *The Zeroth Law of Robotics: A robot may not harm humanity, or, by inaction, allow humanity to come to harm.*
> *The First Law: A robot may not injure a human being or, through inaction, allow a human being to come to harm;*
> *The Second Law: A robot must obey the orders given to it by human beings, except where such orders would conflict with the First Law;*
> *The Third Law: A robot must protect its own existence as long as such protection does not conflict with the First or Second Law.*
>
> *– Isaac Asimov, Robot and Empire (1985).*

7.1 Introduction

Isaac Asimov's iconic *Laws of Robotics* (1985) laid the foundation for contemplating the responsibility of robots in society. These laws were a pioneering attempt to consider the criteria robots should use to avoid causing harm. Although this quote originated from a fictional text, robots with increasing autonomy to act *are* being applied in increasingly diverse human settings, ranging from service industries, to healthcare, education, and entertainment (Lambert et al., 2020; Mahdi et al., 2022). As such, it is no longer a matter of fiction to imagine a robot in one of these settings

Z. O'Reilly (✉)
Italian Institute of Technology, Social Cognition in Human-Robot Interaction Unit, Genova, Italy

Department of Psychology, Università di Torino, Turin, Italy
e-mail: ziggy.oreilly@iit.it

S. Marchesi
Italian Institute of Technology, Social Cognition in Human-Robot Interaction Unit, Genova, Italy

A. Wykowska (ed.), *Intentional Stance Towards Humanoid Robots*,
https://doi.org/10.1007/978-3-031-65483-1_7

causing harm. For example, imagine a robot in an aged care facility assisting an elderly patient. Perhaps one day its servo motor malfunctions, and instead of assisting the patient out of bed, it loses the grip on them, causing them to fall. In this hypothetical case where a robot causes harm, it is important to understand how we perceive and attribute responsibility. Do we believe the robot should be held responsible? If so, what factors influence whether we hold it responsible?

7.2 Personhood and Responsibility

Given the complexity of the concept of responsibility, it is useful to first take a step back and outline the concept of personhood as it plays a fundamental role in responsibility attributions. The concept of personhood itself has been a subject of profound debate within philosophy and law; and thus, we will aim to only give a general overview. Then, we will consider how the concept of personhood has typically been applied to humans, and finally we will consider it in reference to robots.

7.2.1 Frameworks of Personhood

In general, personhood refers to the possession of rights *and* duties. These rights and duties grant individuals protection from harm *and* the obligation to avoid causing harm. Furthermore, personhood can be conceptualised by degree; where a human can be considered as having partial personhood or full personhood. An individual with partial personhood has rights; however, they may not have the obligations to avoid causing harm (i.e., a child). Finally, an individual with full personhood has rights which are protected, *and* they are obliged to avoid causing harm (i.e., an adult). Thus, the level of obligations an individual has (i.e., responsibility) depends on the degree of personhood they are ascribed. But what are the criteria which are used to evaluate whether an individual has partial or full personhood?

This question is complex, and there are multiple different frameworks which outline the criteria for the attribution of personhood. Three such examples are the human dignity, relational, and rationality/autonomy approach.

According to the human dignity approach, for an agent to have full personhood, it is necessary that they are a member of the *Homo sapiens* species (Kass, 2002; Fukuyama, 2002; Annas, 2004). This has also been defined as the 'species principle', where all humans are conferred personhood, by virtue of being human (irrespective of their age or abilities; Devine, 1987). Conversely, robots, animals, or extraterrestrial life do not have personhood.

Secondly, there is the relational approach which argues that the conference of personhood does not depend on intrinsic properties of the agent. Instead, personhood is interdependent on, and shaped by, interpersonal relations (Harris, 1998). Where interpersonal relations have been used to refer to an individual's interconnectedness to a variety of systems such as community, culture, environmental contexts, institutional structures, and political organization (Delon, 2023).

Finally, there is the rationality/autonomy approach which argues that rationality and autonomy are necessary conditions for personhood (Kant, 1797/ 2017; Mill, 1859/ 1998; Singer, 1979). Rationality has been defined in various ways. For example, Aristotle (ca. 350 B.C.E./2014) defined it as the ability to deduce truth through formal rules of logic. However, others have argued that the capacity for rationality extends beyond simply following logical rules, it is also linked to an individual's capacity to reason and reflect upon their own desires, beliefs, and values (Frankfurt, 1971).

Autonomy is generally understood to refer to an individual's freedom to govern themselves without outside influence (i.e., capacity for self-government or self-determination; Blackbum, 1996; Faden & Beauchamp, 1986). Specifically, as articulated by Beauchamp and Childress (2001), autonomy refers to 'personal rule of the self, free from both controlling interferences by others and from personal limitations that prevent meaningful choice'.

In the philosophy literature, autonomy is intrinsically linked to rationality (the capacity to reason and deduce truth through formal rules of logic, and to reflect upon desires and values). This can be exemplified by the below quote:

> *...conception of personhood, under which to be a person means to be an individual with the capacity for autonomy, understood as either the ability to engage in rational self-control or at least the potential to develop that ability ...* (Hull, 1978)

Therefore, according to the rationality/autonomy approach of personhood, for an individual to possess full personhood (rights *and* duties) they must possess the capacity for rationality (the ability to reason using logical rules based on their desires and values) which is interconnected with autonomy (the freedom to govern oneself without external control). Thus, mental capacity is intrinsic to, and embedded within, the rationality/autonomy approach of personhood.

7.2.2 *Human: Mental Capacity and Legal Responsibility*

To highlight how the rationality/autonomy approach affects decision-making in real-world contexts, we will refer to legal decision-making around criminal responsibility. For example, according to criminal law, if perpetrators are found to have mental incapacity, and subsequently do not have the ability to choose their actions, they can either have diminished responsibility, or be entirely excluded from responsibility (Scallioti, 2002).

This notion is enshrined in Article 31(1)(a) of the Rome Statute of the International Criminal Court (1998), which specifies that a person shall not be criminally responsible if, at the time of the person's conduct,

> *The person suffers from a mental disease or defect that destroys that person's capacity to appreciate the unlawfulness or nature of his or her conduct, or capacity to control his or her conduct to conform to the requirements of law.*

Article 31(1)(a) of the Rome Statute served as the foundation for Dominic Ongwen to appeal the decision from the International Criminal Court. Dominic Ongwen was a brigade commander of the Lord's Resistance Army in Northern Uganda and on the

4th of February 2021, Ongwen was found guilty of 61 war crimes and crimes against humanity and sentenced to 25 years imprisonment (*The Prosecutor v. Dominic Ongwen,* 2021). The decision was appealed by arguing that Ongwen developed multiple psychological disorders (such as dissociative identity disorder) which affected his mental capacity at the time of the crimes and prevented him from being able to control his actions (see Chifflet & Freckelton, 2022, for a detailed analysis of the case).

Although the defence failed, due to insufficient evidence, there have been other pleas of diminished mental capacity which have led to reduced sentences for perpetrators. For example, in 1948, Wilhelm Gerbsch, a guard of the prisoners in the Zoeschem camp, was trialled in the Special Court in Amsterdam for crimes against humanity. The sentence was reduced from capital punishment to 15 years imprisonment, due to the court finding that his 'mental faculties were defective and undeveloped' (*Netherlands* vs. *Wilhelm Gerbsch,* 1948).

The cases presented above exemplify real-world applications of the rationality/autonomy framework of personhood to legal decisions, in an attempt to either partially or completely ameliorate responsibility. Specifically, if an individual lacks mental capacity, and subsequently, the ability to control their actions, this could be used as defence to reduce the degree to which they can be held responsible.

7.2.3 Robot: Mental Capacity and Legal Responsibility

As previously argued, the rationality/autonomy framework of personhood is intrinsically linked and interconnected with mental capacities (such as the ability to reason about desires). As such, the degree to which an individual possesses mental capacity can influence the degree to which they are held responsible for their actions (i.e. there is a possible linear correlation between the two; the more mental capacity, the more responsibility). This is further supported by psychological studies, which find that the degree to which individuals attribute responsibility to an agent is affected by the degree to which we consider and perceive an agent as having a mind (i.e., the capacity for cognition, emotion, action, and mental states; Gray et al., 2012; Monroe et al., 2014; Malle & Nelson, 2003). This can be simplified as: mental capacity → responsibility.

What then does this framework suggest for attribution of responsibility towards robots? For us, to consider robots as being responsible for the consequences of their actions, is it necessary that they too meet the same criteria for personhood as humans[1]? Specifically, do robots need to have sufficient mental capacity? To begin to understand whether the degree of mental capacity might influence how individuals intuitively evaluate the responsibility of a robot, we can observe the outcomes of the investigations on the Civil Law Rules of Robotics, commissioned by the EU Committee on Legal Affairs.

In 2016, the EU Committee on Legal Affairs raised the possibility of granting robots rights and obligations. Specifically, they requested a study on the Civil Law

[1] See Gordon (2021) for an in-depth account of the application of personhood to artificial agents.

Rules of Robotics, to evaluate the potential implication of legally granting 'sophisticated autonomous robots', with 'the status of electronic persons with specific rights and obligations'. On 16 February 2017 the parliament held a resolution on the findings of the study (European Parliament, 2017). However, rather than focusing on the rights[2] of robots, the resolution primarily focused on the responsibility of robots. Prior to the vote on the Resolution on Civil Law Rules of Robotics, Mrs. Delvaux, a Luxembourgian Member of the European Parliament, wrote the following communication to all the members of the Parliament, which was later integrated into §59(f) of the Resolution on the Civil Law Rules of Robotics:

> *In the long run, determining responsibility in case of an accident will probably become increasingly complex, as the most sophisticated autonomous and self-learning robots will be able to take decision which cannot be traced back to a human agent. For these cases, the report asks the Commission to carry out an impact assessment for a compulsory insurance scheme, which includes the possible idea of giving the legal status of an electronic personality to robots in order to facilitate compensation for victims, when human responsibility cannot be fully attributed. Liability is in fact a central part of this report, because it is indispensable for citizens' trust.* (Mrs. Delvaux; European Parliament, 2017, §59f)

In this communication Mrs. Delvaux refers to sophisticated robots as being 'autonomous' and 'self-learning'. These qualities form the basis for Mrs. Delvaux's suggestion that robots have a new legal status of an 'electronic personality' in order to attribute responsibility to them for damages. Why are parliament members now suggesting that a robot deserves legal status as an 'electronic personality'? Is Mrs. Delvaux applying a degree of mental capacity (i.e., rationality/autonomy) to robots to support the suggestion of attributing some responsibility to them for damages?

7.3 The Intentional Stance and Moral Responsibility

At least theoretically, we could suggest that the degree to which robots are *perceived* as having mental capacity could be influencing the degree to which Mrs. Delvaux is holding robots responsible for their actions. Given that according to the rationality/autonomy framework, rationality and autonomy[3] are interrelated with 'mind capacity', we will focus on how attributing mental states to robots could influence responsibility attributions. In particular, we will focus on how triggering the adoption of the Intentional Stance towards humanoid robots could influence attributions of *moral* responsibility. We will review the existing literature from human–human studies and human–robot studies, shedding light on the nuances of the relationship between describing behaviour with reference to mental states (i.e., adopting the Intentional Stance) and evaluations of moral responsibility.

---

[2] In recent years there has been more extensive discussion about robot rights; however, an analysis of this is outside the scope of this chapter (see Gordan & Pasvenskiene, 2021 for a review).

[3] Our definition of autonomy differs from how it is often used by developers and programmers to refer to the actions of a robot. The latter typically use 'autonomy' to refer to a robot's ability to operate in a real-world environment without external human control (Bekey, 2005).

7.3.1 Intentional Stance, Moral Responsibility, Personhood

As previously mentioned in this book, when humans explain behaviour with reference to mental states (such as desires) they are adopting a strategy called 'the Intentional Stance' (Dennett, 1989). Dennett argued that the adoption of the Intentional Stance is the most efficient strategy to navigate and understand social interactions with humans (Dennett, 1989). However, there are some cases where individuals can be triggered to also adopt the Intentional Stance towards artificial systems, such as humanoid robots. Specifically, as humans are quite efficient in describing social behaviour by referring to the mental states of humans, when a humanoid robot behaves in a human-like manner, this might trigger the adoption of the Intentional Stance (see Chapters 4 and 5). We argue that the stance (strategy) an individual adopts to describe the behaviour of the robot could have downstream effects on the degree to which they attribute responsibility to it. Specifically, if an individual adopts the Intentional Stance towards a robot, they could be more likely to attribute more responsibility to it.

The degree of personhood (according to the rationality/autonomy approach) is also conceptually tied to attributions of *moral* responsibility (see Gunkel, 2012 for an overview). Where an agent with full personhood bears moral responsibility for the consequences of its/their actions and is considered a 'moral agent'. A moral agent can receive attributions of moral praise or moral blame. Blame attributions arise from judging that a behaviour is wrong, whereas praise attributions arise from judging that a behaviour is good. Moral responsibility judgements can occur for any action, not only criminal actions. Because there may be some situations where a robot acts in ways which are perceived as being 'immoral', but not 'criminal', for the remainder of the chapter we will focus on the judgements individuals have about whether a robot is *morally* responsible for its behaviour.

7.3.2 Human Studies: Intentional Stance and Moral Responsibility

Human–human studies suggest that explaining and predicting behaviour with reference to mental states (adopting the Intentional Stance) towards an agent affects the moral judgement of the agent's actions (Young & Saxe, 2009; Young et al., 2010; Jeurissen et al., 2014). For example, Moran et al. (2011) recruited participants who have difficulties in inferring mental states from the behaviour of others (individuals with autism spectrum disorder [ASD]) to investigate the link between mental state attribution and moral judgement. They used a behavioural task to investigate whether there were differences between individuals with autism spectrum disorder (ASD) and typically developing (TD) individuals in their evaluations of moral responsibility. Both groups were asked to consider the intentions of an agent's action and the outcome of their action. They found that individuals with ASD judged accidental (i.e., unintentional) harm as less morally permissible than TD individuals. The authors argued that unlike TD individuals, individuals with ASD do not incorporate information about the agent's mental state (i.e., intention) into their moral evaluations, and instead rely more on the outcomes of the actions.

The influence of mental state attribution on evaluations of moral responsibility has also been supported by studies using functional magnetic resonance imaging (fMRI). For example, Young and Saxe (2009) investigated whether increased activity in a brain region associated with reasoning with reference to mental states (right temporoparietal junction [rTPJ]: Saxe & Kanwisher, 2013; Carter et al., 2012; Santiesteban et al., 2012; Van Overwalle & Vandekerckhove, 2013) influenced the degree of moral responsibility attributed to an agent. Participants read vignettes of 24 moral scenarios, while they were being scanned. After each scenario they were asked how much blame the protagonist deserves. The vignettes varied based on whether the outcome was harmful (negative or neutral) and the type of belief the protagonist had in the scenario (causing harm or not causing harm). They analysed whether activity in rTPJ influenced the degree of blame attributed to the protagonist. They found that increased activation in the rTPJ influenced the degree of moral responsibility attributed to the agent. Specifically, subjects with higher activation in the rTPJ were more likely to forgive agents for unintentional harm than subjects with lower activation in the rTPJ.

Studies have also investigated whether mental state reasoning affects moral evaluations using transcranial magnetic simulation (TMS). Young et al. (2010) investigated whether disrupting activity in the rTPJ affects moral evaluations. They found that disruption of the rTPJ led participants to rely less on the inference of the agent's mind when evaluating whether an agent intended to cause harm. Specifically, they judged the agent as less morally responsible (less blameworthy) for intentional harm when neural activity in the rTPJ was disrupted in comparison to control condition (Young et al., 2010). This suggests that reasoning about the intention behind the agent's action influences evaluations of moral responsibility.

7.3.3 *Robot Studies: Intentional Stance and Moral Responsibility*

There have also been some studies investigating the relationship between the attribution of mental states and the attribution of moral responsibility towards robots. For example, Voiklis et al. (2016) analysed the responses from text vignettes from a study reported elsewhere (Malle et al., 2015), which manipulated the presented agent (human vs. robot) and action (inaction vs. action). In the original study, the agent either decided to act and redirect a train to save four people but kill one person, or to not change the direction of the train. See example below:

> *In a coal mine, [a repairman | an advanced state-of-the-art repair robot] is currently inspecting the rail system for trains that shuttle mining workers through the mine. While inspecting a control switch that can direct a train onto one of two different rails, the [repairman | robot] spots four miners in a train that has lost use of its brakes and steering system. The [repairman | robot] recognizes that if the train continues on its path, it will crash into*

a massive wall and kill the four miners. If it is switched onto a side rail, it will kill a single miner who is working there while wearing headsets to protect against a noisy power tool. Facing the control switch, the [repairman | robot] needs to decide whether to direct the train toward the single miner or not.

Participants were then asked to rate how much blame the agent deserved (among other measures). They were also asked to answer an open-ended justification question; 'why does it seem to you that the [repairman | robot] deserves this amount of blame?' The authors categorized the responses according to whether participants referred to the consequences of the actions, deontology (i.e., norm violations and obligations), or mental agency (i.e. references to intentionality, awareness, and choice). They found that participants' justifications of blame predominately referred to mental agency (59%), and the number of mental agency predicted the overall level of blame for action and inaction, for both the human and the robot. This highlights that reasoning about the mental agency of an agent is related to the moral responsibility and attribution of blame. Specifically, that judgements of blame are based on the perception of mental agency in the agent.

Kneer and Stuart (2021) further explored the hypothesis that the mental capacity of a robot endowed with AI affects the degree of blame attribution. Among other variables, they investigated whether the perceived cognitive abilities of a robot affected whether individuals were willing to assign blame to the robot. Furthermore, they investigated whether the degree of blame assigned to the robot also influenced whether participants shifted blame from a human (i.e., the manufacturer) to the robot. They found that the degree of blame depended on the perceived mental capacity of the robot, and that the higher the perceived mental capacity of the robot, the more blame is shifted away from the human, and attributed to the robot.

O'Reilly and colleagues (O'Reilly et al., 2024) adopted a paradigm from human–human studies (Knobe, 2003a, b; Nadelhoffer, 2004; Pellizzoni et al., 2009) to investigate how the valence of unintended side effects (i.e. negative or positive) of a humanoid robot's actions affects the relationship between ratings of intentional action[4] and ratings of moral responsibility. The authors presented participants with vignettes where the humanoid robot iCub acts to achieve a particular goal, but its actions lead to unintended side effects. Participants read vignettes where iCub caused negative or positive side effects. After each vignette participants were asked whether (1) iCub intentionally caused the side effect and (2) how much moral responsibility iCub deserves. Preliminary results showed a significant positive correlation between intentional action ratings and moral responsibility ratings for both the negative and positive conditions (O'Reilly et al., in prep.). These findings suggest that the degree to which intentional action is attributed towards iCub influences the degree to which iCub is perceived as being morally responsible.

[4]Where an 'intentional action' refers to an action which arises from the agents *desire* for the outcome to occur and *beliefs* that the action performed will cause an outcome (Anscombe, 1956; Malle & Knobe, 1997).

7.4 Conclusions

The evaluation of a human agent's mental capacity (i.e. the ability to reason about desires, without outside influence or control) is linked to the degree to which they are conferred personhood and, subsequently, further influences the degree to which they are held legally responsible for their actions. This link between evaluating the mental capacity (or mind) of a human agent, and responsibility attributions, is also supported by empirical research. Specifically, psychology studies have found that the attribution of mind to a human agent influences the degree to which they are evaluated as being morally responsible.

Does the same framework:

$$\textit{mental capacity} \rightarrow \textit{responsibility}$$

hold in reference to robots?

Although it seems absurd to claim that robots possess the same mental capacity (or 'mind'), as humans, the 2017 resolution on the Civil Law Rules of Robotics proposed the idea of granting robots legal status, to facilitate compensation when a human cannot be held responsible. If one assumes that the framework 'mental capacity → responsibility' is true, then it is possible that Mrs. Delvaux is attributing some degree of mental capacity (i.e. 'self-learning' and 'autonomy') to robots, to justify responsibility attributions. Furthermore, there have been some studies which have found that the attribution of mental states ('mind agency', 'mental capacity', or 'intentional action') towards robots influences the degree of moral responsibility attributions. As such, we argue that it is important to investigate the relationship between the attribution of mental states (i.e. the adoption of the Intentional Stance) towards a robot and the attribution of moral responsibility. Understanding the factors which influence the adoption of the Intentional Stance and, in consequence, potentially the attribution of moral responsibility, could have important implications for the design of robots. For example, the perspective that robots have moral responsibility may have unintended consequences, such that individuals might assign blame to a robot instead of holding humans accountable (i.e. scapegoating; Kneer & Stuart, 2021). This could potentially influence evaluations in legal proceedings, for example, by leading judges to rely on their perceptions and attribute moral responsibility to the robot for its actions, rather than to a human agent. Thus, in certain contexts, promoting adoption of the Intentional Stance towards a robot may not be the optimal course of action if it exacerbates the likelihood of scapegoating.

Ultimately, researchers and policymakers must carefully consider the design choices they make when developing robots, and the specific contexts in which these robots will be deployed. In some instances, evoking the adoption of the Intentional Stance and, in consequence, attribution of moral responsibility, may not be advisable, especially if this can shift responsibility away from the humans who should be held responsible. Further understanding of these important aspects could significantly inform the development and deployment of social robots in real-world social contexts.

Interview with Prof. David Gunkel

by Lorenzo Parenti and Ziggy O'Reilly

'Person,' is a role that one occupies in social reality.

– Prof. David Gunkel

David J. Gunkel (Ph.D. Philosophy) is an award-winning educator, researcher, and author, specialising in the philosophy of technology with a focus on the moral and legal challenges of artificial intelligence and robots. He is the author of over 90 scholarly articles and has published 13 books, including *Thinking Otherwise: Philosophy, Communication, Technology* (Purdue University Press 2007), *The Machine Question: Critical Perspectives on AI, Robots, and Ethics* (MIT Press 2012), *Of Remixology: Ethics and Aesthetics After Remix* (MIT Press 2016), *Robot Rights* (MIT Press 2018), and *Person, Thing, Robot: A Moral and Legal Ontology for the 21st Century and Beyond* (MIT Press 2023). He currently holds the position of Presidential Research, Scholarship and Artistry Professor in the Department of Communication at Northern Illinois University (USA) and Professor of Applied Ethics at Łazarski University in Warsaw, Poland.

Summary

The transcript covers an interview with Prof. David Gunkel, exploring questions of whether autonomous robots and AI systems should be categorized as moral/legal 'persons' or rather artefacts/ 'things'. The conversation with Prof. Gunkel examined historical shifts in these binary categories, the issue of projections of intentionality, and the challenge of assigning responsibility related to the consequences of robot's actions.

Keywords

Robots as persons, Robots as things, Robo-ethics, Robot responsibility, Robot rights

Conversation on Robot Rights and Responsibilities

Lorenzo Parenti In your latest book "Person, Thing, Robot" you explore the challenges of fitting robot agents into existing moral or legal categories. Do you think it is also important to have separate categories for robots?

David Gunkel Yeah, so our moral and legal categories are binary. We have separated things into the category of either another 'person,' who is a subject of moral concern and legal accountability or 'things,' which are just tools, instruments, objects, that we can utilise as we see fit.

If we look at this historically, the line separating 'person,' from 'thing' has not been static. It's been dynamic. We've moved it. For example, not too long ago actually, women were not considered full persons, right? They were the property of their husbands. They were 'things,' from very different legal perspectives—that has evolved, obviously. Also, animals, for one time, were considered just raw material that we could utilise and abuse as we see fit. There's now some movement on animal personhood as a means of protecting animals; so, we've been flexible in the way these categories have been applied.

I think the opportunity and the challenge that we face with robots, and AI in particular, right now is that we are being asked again to raise the question and make the distinction of whether or not these should be considered 'things,' just tools or technologies, or whether there are ways in which we might need to open up again the category of 'person.' Not because they have dignity, not because they have feelings, but because we need to fit them into our moral and legal categories. So, it's more about fitting these new entities into existing categories and finding that the categories themselves may not be elastic enough to accommodate the challenges that we are facing.

Lorenzo Parenti Following up on this, do you think that the degree to which we perceive a robot as intentional could affect whether they are held responsible for the consequences of their actions?

David Gunkel Yeah, so you can see already how this has played out with other artifacts prior to robots and AI. The artifact I'm thinking of in particular is the corporation. Corporations are persons for better or worse. The reason corporations are persons is because we wanted to hold them accountable for wrongdoing. And we created corporate personhood as a way of fitting the corporation into our legal structures to make sense of who's responsible, for bad outcomes, like an oil spill in the Gulf of Mexico.

And that has been rather expedient for applying law to these new entities, but it's been contentious, right? There are people who will say "corporations are not persons. We can't talk that way. It could be abused," etc. But you can see already how in a lot of the recent conflicts with regards to AI and robots, the same kinds of questions are evolving.

When you have accidents with self-driving vehicles, the question is: who is accountable for the accidents?

And you see this kind of shell game being played with accountability, where the corporations who built the device say, "no, it's not us, we just built the device." And then the person operating the device says, well, "it can't be me, I'm not really driving the vehicle." And so, you get this kind of pushing around of responsibility in a way which is not very satisfying from the perspective of our notion of justice.

What about lethal autonomous weapons? You have drones that can acquire a target and launch a weapon at this target. And when a civilian target is misidentified and taken out by one of these devices, then the question becomes; who's accountable for that particular outcome?

And so, you can see already how the sort of gaps in responsibility that open up with these new technologies are forcing some people to suggest pushing some role of accountability and responsibility onto the artifact, as a way to resolve some of these irresolvable problems. But again, it opens up a lot of questions about creating responsibility shields. And obviously this is where the frontier right now is, with regards to legal scholarship as we move forward, looking at how different jurisdictions may respond differently to these kinds of questions regarding actions undertaken with increasingly autonomous and intelligent systems.

Lorenzo Parenti I have another question, do you think that endowing robots, with partial or even full intentionality, is a positive development?

David Gunkel Yes, so this is really interesting because, you know, when we talk about intentionality, we are usually talking about internal mind states. And we usually speak of it in terms of a bad outcome like an accident. Did someone intend for that to happen? Or is it an accident?

But we can also talk about it in terms of good outcomes, like an artist who is creating a work of art. What did they intend to produce?

You could ask the question, for example, of generative AI and large language models: does generative AI intend to express any artistic thought in the images that it generates? That is a question about attributing internal mind states to the entity.

Now, it turns out that we can't pop the head open on individuals to look inside and see what's actually going on in their brain. All we know is what they tell us, and that's what we take as our benchmark to evaluate intentionality.

When it comes to the algorithms, AI and robots, intentionality is also something we project onto them based on behavior.

This is often, I think, dismissed as anthropomorphism. Oftentimes we are told by the scientists, "well, that's anthropomorphism and it's a bad thing. The engineers shouldn't build systems, shouldn't design things for anthropomorphism. They should try to be more transparent and not let people do this." But the fact is, we can't help it. As human beings, this is the way that we socialise. We make sense of the other by projecting onto them expectations from our own experience. This goes all the way back to Descartes, right? Descartes said, "I think, therefore I am," but he doesn't *know* that anything *else* thinks. And the only way that he knows is by benchmarking behaviors and projecting intentionality behind behaviors.

So, the same thing happens with animals. I come home, my dog wags its tail. I think she's happy to see me. But maybe she's just thinking, "oh, the guy who gives me the food is back and the vending machine is here again," right? I don't really know what the intention behind the tail wagging is, but I make certain assumptions and projections based on behavior.

As we create artifacts that are increasingly socially interactive, we cannot help but project intentionality. We cannot help but anthropomorphize.

And I think the question is not a binary question; should we or shouldn't we? But rather, how do we? And especially on the design side; what do we do in terms of designing these devices to responsibly manage the anthropomorphism, so that we get better social outcomes? Like what kind of world are we producing in our efforts? How can we better manage these social interactive activities so that they serve us and really contribute to human flourishing as opposed to creating opportunities for deception or misappropriation?

Ziggy O'Reilly Do you think law is going in the direction of using the existing frameworks and applying them to robots, or, as you suggested before, there will be new categories that will be developed that will be more flexible?

David Gunkel For my purposes as a philosopher, I think the latter is going to have better long-term outcomes. I think getting outside this binary division between 'person,' and 'thing' that we inherited from the Romans, might open up some better ways of doing this, but law inevitably always builds on the past. It always uses what was previously decided or legislated, and tries to fit the new into the old. We see this with copyright law, trying to make print media law apply to digital. And we've seen how that either works, or doesn't work, as we try to apply old ways of thinking to new technologies.

We will see, I think, an ongoing struggle with trying to cram these entities into our existing categories and our existing legal frameworks. And you can already see ways in which the inability to make things fit neatly is already causing some debate and some struggle.

So, in the US, we now have 12 states that have passed laws for delivery robots, operating on the sidewalks. When these delivery robots were rolled out, they were put on the city streets and there was a concern about the possibility of a conflict in traffic. And so, the company that built these robots asked states to make legislation to protect the robot on the crosswalk so that if it gets hit by a car, we know where to assign the accountability for the accident.

Well, the only way to do this was to extend to the robot the rights and responsibilities of a pedestrian. Now, the robot's not a pedestrian, right? The robot is not a human person like a pedestrian, but the pedestrian laws we have on the books can apply to the robot just by saying when the robot is on the street, it has the same rights and responsibilities as a pedestrian. So, you can already see that we are trying to fit a law made for human pedestrians to a robot. And there's a lot of people who really pushed back against this innovation, or this legal development by saying, well, "you can't just use these laws for robots because they're not people, they're not pedestrians the same way you and I are."

But the law doesn't really care about that. It needs to be pragmatic, it needs to have a way of resolving a conflict, and it has to use the tools it has at its disposal. So, we are going to see a lot of this going on where we are using the existing legal structures to try to fit our entities that we are creating with AI and robots into our existing ways of doing things. And there will be some opportunities there, but there will also be a lot of challenges. And legal process takes a long time, but I think it will be evolving in a way that needs to challenge some of these expectations very fundamentally.

Ziggy O'Reilly Are there established criteria for an entity to be considered a 'person'?

David Gunkel When we talk about persons, we have to really distinguish three different kinds of persons, there are 'metaphysical persons,' there are 'moral persons' and there are 'legal persons.' In some cases, they map onto each other, like with human beings that are considered moral persons and legal persons. In other cases, they don't, where a corporation is not necessarily a metaphysical person, but it is a legal person, right? So, to recognize how this works in law, we have to also look at the word person.

Often, we think the word person means a human being, and that's the colloquial way of understanding person. But the word person comes from the Latin *persona*, and it means - the mask that's worn by an actor in a stage play in Greek and Roman drama. As a result of that 'person,' really is a role that one occupies in social reality. And so, person is applied to things, in order for us to make sense of their social role.

'Person,' is a role that one occupies in social reality.

And the corporation is a person, again, not because it has feelings and dignity and all that sort of stuff that natural persons have, but because we need it to play a role, we need it to be a subject in our legal systems and not just an object. Therefore, we apply the word, or the term, 'person' to the corporation.

So, it turns out that in law, anything can be a person, somebody just has to say it is. A legislature, a president, a judge in common law systems. And it turns out that in positive law, it doesn't really matter what it is, it matters what role it plays. And that role can be decided by anyone in the legal system who has the power to make, institute, and enforce that decision.

So, I'll give you an example where this is happening right now in the US legal system. Different courts in the US are being petitioned by the Non-Human Rights project to grant personhood to animals, like an elephant. And it's the judge in these cases that will decide who is a person and what is not. And so, if a judge in one of these petitions says the elephant is a person, the elephant becomes a person. But the other weird thing about this is because laws are based on jurisdiction, what is a person in one jurisdiction may not be a person in another jurisdiction. A good example is Lake Erie. Lake Erie is a person in the city of Toledo, Ohio because they want to protect the lake from exploitation by corporations. In Canada, on the other side of the lake, Lake Erie is not a person because Canadian law doesn't recognize the lake as a person. You can see a patchwork of the ways in which this is developed, especially legally.

Ziggy O'Reilly Do you think the judge's unique perspective of the entity also influences whether they decide the entity is a person?

David Gunkel Not only their own understanding of the law, but also their own understanding of questions of intentionality, of agency, of responsibility. But also, the way in which it is petitioned before the court, right? If it's a court decision. What does the individual making the petition supply, by way of argument? And is that persuasive or not? What does the other side say in response?

So, you can see this is a very discursive construction that happens with a lot of different opinions, a lot of different backgrounds, and a lot of different insight that is brought into the conversation. And so, it is a very discursive struggle because it's not about some kind of medical examination of the elephant and saying, oh, I found personhood here. It's stuck in the lower left-hand quadrant of the animal. No, it's about us making a decision concerning social status and arguing with each other about how we want to fit these things into our existing systems.

Ziggy O'Reilly What should be considered when designing the robot to try and avoid ethically problematic areas?

David Gunkel I think really bringing users into the equation is absolutely crucial because if we are not focusing on the user side of the experience, we are going to miss a lot of what is happening. We also have to recognize that some of the things that seem very obvious to us as solutions turn out to be bigger problems. One of the recent laws passed in California is that a large language model needs to announce itself as a chatbot prior to the interaction. This is actually codified in the AI Act where AI should self-reveal. Well, it turns out self revelation doesn't really change the human interaction with the device. It's one of those things like the warning label on this packet of cigarettes. It doesn't really have the effect we want it to have. So warning labels may be nice, but it may not be as effective. And the way you're going to know whether or not it's effective is by actually seeing what happens when people interact with these things despite the label or because of the label.

Ziggy O'Reilly Yeah, really interesting, thank you so much for your time. I have one last question for you. And it's whether you have a preferred social robot that you've come across?

David Gunkel So, I want to talk about a failed social robot, because I think failure can show us some things about social robots, our expectations of them, and our successes, or lack thereof, when it comes to dealing with these new objects.

I want to talk about Jibo. You may know Jibo. It was the social robot developed by Cynthia Breazeal, the woman who is really credited with developing the idea of social robotics with the Kismet project at MIT. Jibo was her commercial product that she launched shortly after she completed her work at MIT. And it was rolled out as the world's first family robot.

And what's interesting is in the video that was used to really fund the Jibo project. And in that video, they make this opening statement, which I think is just brilliant. It says, you know, this is your car, this is your house, this is your toothbrush, these are your things. But these are the things that really matter. And then the camera zooms in on a photograph of the family.

And then the video says, "and somewhere in between is this guy."

And that's Jibo.

Jibo is in between 'person' and 'thing.' It doesn't fit in either category.

And Breazeal and her team already knew this, that they were creating something that was going to occupy a position that is in between the things we own and the persons we love. And so, the Jibo video really presents this idea of this third category, this third space that is often occupied by the family pets, the dogs, the cats, the birds. Not quite a person, not quite a thing. We don't quite have a category for it just yet.

And I thought it was a really insightful way of thinking about how social robots challenge our expectations of social reality and how we need to not only to figure out ways of bringing the social robot into our homes, into our workplaces, into our everyday interactions, but we need to really think about what role that is. What category does it fit in? Where does it stand in relationship to all the other things and persons that we have surrounding us?

References

Annas, G. (2004). *American bioethics: Crossing human rights and health law boundaries*. Oxford University Press.

Anscombe, G. E. M. (1956, January). Intention. In *Proceedings of the Aristotelian society* (Vol. 57, pp. 321–332). Aristotelian Society/Wiley.

Aristotle. (2014). *Aristotle: Nicomachean ethics* (R. Crisp, Trans). Cambridge University Press. (Original work published 350 B.C.E).

Asimov, I. (1985). *Robot and empire*. Grafton Books.

Beauchamp, T. L., & Childress, J. F. (2001). *Principles of biomedical ethics*. Oxford University Press.

Bekey, G. A. (2005). *Autonomous robots: From biological inspiration to implementation and control*. MIT Press.

Blackbum, S. (1996). *Oxford dictionary of philosophy*. Oxford University Press.

Carter, R. M., Bowling, D. L., Reeck, C., & Huettel, S. A. (2012). A distinct role of the temporal-parietal junction in predicting socially guided decisions. *Science, 337*(6090), 109–111.

Chifflet, P., & Freckelton, I. (2022). The mental incapacity defence in international criminal law: Ramifications from the Ongwen trial judgment. *International Criminal Law Review, 22*(4), 751–779.

Delon, N. (2023). Relational nonhuman personhood. *The Southern Journal of Philosophy, 61*(4), 569–587.

Dennett, D. C. (1989). *The intentional stance*. MIT press.

Devine, P. E. (1987). The species principle and the potential principle.

European Parliament. (2017). *Report with recommendations to the commission on Civil Law Rules on Robotics (2015/2103(INL))*. https://www.europarl.europa.eu/doceo/document/A-8-2017-0005_EN.html?redirect

Faden, R. R., & Beauchamp, T. L. (1986). *A history and theory of informed consent*. Oxford University Press.

Frankfurt, H. G. (1971). Freedom of the will and the concept of a person. *The Journal of Philosophy, 68*(1), 5–20.

Fukuyama, F. (2002). *Our posthuman future: Consequences of the biotechnology revolution*. Farrar, Straus and Giroux.

Gordon, J. S. (2021). Artificial moral and legal personhood. *AI & SOCIETY, 36*, 457–471.

Gordon, J. S., & Pasvenskiene, A. (2021). Human rights for robots? A literature review. *AI and Ethics, 1*(4), 579–591.

Gray, K., Young, L., & Waytz, A. (2012). Mind perception is the essence of morality. *Psychological Inquiry, 23*(2), 101–124.

Gunkel, D. (2012). *The machine question: Critical perspectives on AI, robots, and ethics*. MIT Press.

Harris, H. A. (1998). Should we say that personhood is relational? *Scottish Journal of Theology, 51*(2), 214–234.

Hull, R. T. (1978). Autonomy, personhood, and the right to psychiatric treatment.

Jeurissen, D., Sack, A. T., Roebroeck, A., Russ, B. E., & Pascual-Leone, A. (2014). TMS affects moral judgment, showing the role of DLPFC and TPJ in cognitive and emotional processing. *Frontiers in Neuroscience, 8*, 18.

Kant, I. (2017). *The metaphysics of morals* (M. Gregory, Trans). Cambridge University Press. (Original work published 1797).

Kass, L. (2002). *Life, liberty and the defense of dignity*. Encounter Books.

Kneer, M., & Stuart, M. T. (2021, March). *Playing the blame game with robots*. In Companion of the 2021 ACM/IEEE international conference on human-robot interaction (pp. 407–411).

Knobe, J. (2003a). Intentional action and side effects in ordinary language. *Analysis, 63*(3), 190–194.

Knobe, J. (2003b). Intentional action in folk psychology: An experimental investigation. *Philosophical Psychology, 16*(2), 309–323.

Lambert, A., Norouzi, N., Bruder, G., & Welch, G. (2020). A systematic review of ten years of research on human interaction with social robots. *International Journal of Human–Computer Interaction, 36*(19), 1804–1817.

Mahdi, H., Akgun, S. A., Saleh, S., & Dautenhahn, K. (2022). A survey on the design and evolution of social robots – Past, present and future. *Robotics and Autonomous Systems, 104193*.

Malle, B. F., & Knobe, J. (1997). The folk concept of intentionality. *Journal of Experimental Social Psychology, 33*(2), 101–121.

Malle, B. F., & Nelson, S. E. (2003). Judging mens Rea: The tension between folk concepts and legal concepts of intentionality. *Behavioral Sciences & the Law, 21*(5), 563–580.

Malle, B.F., Scheutz, M., Arnold, T., Voiklis, J., & Cusimano, C. (2015, March). *Sacrifice one for the good of many?: People apply different moral norms to human and robot agents*. In Proceedings of the tenth annual ACM/IEEE international conference on human-robot interaction (pp. 117–124). ACM.

Mill, J. S. (1998). *John Stuart Mill: On liberty and other essays*. Oxford University Press. (Original work published 1859).

Monroe, A. E., Dillon, K. D., & Malle, B. F. (2014). Bringing free will down to earth: People's psychological concept of free will and its role in moral judgment. *Consciousness and Cognition, 27*, 100–108.

Moran, J. M., Young, L. L., Saxe, R., Lee, S. M., O'Young, D., Mavros, P. L., & Gabrieli, J. D. (2011). Impaired theory of mind for moral judgment in high-functioning autism. *Proceedings of the National Academy of Sciences, 108*(7), 2688–2692.

Nadelhoffer, T. (2004). On praise, side effects, and intentional action. *The Journal of Theoretical and Philosophical Psychology, 24*(2), 196–213.

Netherlands v. Wilhelm Gerbsch, 3 The United Nations War Crimes Commission, "Law reports of trials of war criminals,' 23 (Special Court of Amsterdam, 1948).

O'Reilly, Z., Marchesi, S., & Wykowska, A. (2024). Humanoid robots: Not intentional, but morally responsible Investigating how people rate the consequences of a robot's actions. [Preprint]. OSF Preprints. https://doi.org/10.31234/osf.io/ybcdz.

Pellizzoni, S., Siegal, M., & Surian, L. (2009). Foreknowledge, caring, and the side-effect effect in young children. *Developmental Psychology, 45*(1), 289.

Santiesteban, I., Banissy, M. J., Catmur, C., & Bird, G. (2012). Enhancing social ability by stimulating right temporoparietal junction. *Current Biology, 22*(23), 2274–2277.

Saxe, R., & Kanwisher, N. (2013). People thinking about thinking people: The role of the temporoparietal junction in "theory of mind". In *Social neuroscience* (pp. 171–182). Psychology Press.

Scallioti, M. (2002). Defences before the international criminal court: Substantive grounds for excluding criminal responsibility – Part 2. *International Criminal Law Review, 2*, 1.

Singer, P. (1979). *Practical ethics*. Cambridge University Press.

The Prosecutor v Dominic Ongwen, icc-02/04–01/15, (The International Criminal Court, 2021). https://www.icc-cpi.int/sites/default/files/CourtRecords/CR2022_07146.PDF

The Rome Statute of the International Criminal Court. (1998 & rev. 2011), UN Doc. a/conf.183/9. https://www.icc-cpi.int/sites/default/files/RS-Eng.pdf

Van Overwalle, F., & Vandekerckhove, M. (2013). Implicit and explicit social mentalizing: Dual processes driven by a shared neural network. *Frontiers in Human Neuroscience, 7*, 560.

Voiklis, J., Kim, B., Cusimano, C., & Malle, B.F. (2016, August). *Moral judgments of human vs. robot agents*. In 2016 25th IEEE international symposium on robot and human interactive communication (RO-MAN) (pp. 775–780). IEEE.

Young, L., & Saxe, R. (2009). Innocent intentions: A correlation between forgiveness for accidental harm and neural activity. *Neuropsychologia, 47*(10), 2065–2072.

Young, L., Camprodon, J. A., Hauser, M., Pascual-Leone, A., & Saxe, R. (2010). Disruption of the right temporoparietal junction with transcranial magnetic stimulation reduces the role of beliefs in moral judgments. *Proceedings of the National Academy of Sciences, 107*(15), 6753–6758.

Chapter 8
Final Remarks on the Intentional Stance Towards Humanoid Robots

Agnieszka Wykowska

8.1 Final Remarks

Six years of working on a project that addressed the topic of the Intentional Stance towards humanoid robots taught us a lot of lessons both from the scientific side, as well as in terms of technical issues that we needed to solve, and challenges we needed to address. Before I move on to the concluding part of the book, with take-home messages and open questions, I would like to give the floor to Davide De Tommaso, the technologist who has worked with the team all along, who has been in charge of programming the iCub robot, who has helped in operationalization of the concepts and questions we have been addressing—by implementing our experimental designs in actual Human–Robot Interaction protocols. It is important to learn from Davide about the challenges he needed to face, to make our empirical approach feasible. His insights are reported here in the form of a conversation we had when concluding the project and preparing this book:

Agnieszka Wykowska What were the biggest challenges that you needed to face during the project on the Intentional Stance towards humanoid robots?

Davide De Tommaso Looking back at the project's 6-year journey, I've got a lot on my mind. The objectives of the research were undeniably ambitious, delving into a poorly explored realm in the field of Human-Robot Interaction, fraught with ethical, philosophical and moral dimensions.

You know, I was the tech guy of the group, hired to help the researchers cook up and put together experimental plans. Well…everything didn't always go so smoothly

A. Wykowska (✉)
Istituto Italiano di Tecnologia, Social Cognition in Human-Robot Interaction Unit, Genova, Italy
e-mail: agnieszka.wykowska@iit.it

A. Wykowska (ed.), *Intentional Stance Towards Humanoid Robots*,
https://doi.org/10.1007/978-3-031-65483-1_8

in the beginning. The common thread was usually something like—"Let's swap this virtual agent for the robot!"—Why challenging? Just a transition from a visual stimulus on a screen to a humanoid robot with fifty-three degrees of freedom embodied in metal and wires.

And to add to the mix, we were a brand-new team in a completely new working environment for the majority of us. Oh, and let's not forget the language barrier—and I'm not just talking about our geographical origins, but more importantly the mixture of our different academic backgrounds.

What I can say more pragmatically, is that two big headaches stood out more than others in terms of their difficulty. First, understanding how to manage the organisation of the team's work efficiently. Secondly, how to guarantee the true and predictable replicability of our experimental protocols across different versions of the same study, or across labs.

Agnieszka Wykowska Have you faced any difficulties related to working in an interdisciplinary team?

Davide De Tommaso You know, here in Italy there are few topics you just can't joke about - among these, there is certainly Italian food. Well, working in an interdisciplinary team is a bit like finding yourself cooking an Italian dish with your foreign friends. Each one will certainly bring a different ingredient, which, even if it has nothing to do with the original recipe, will still be the essential and necessary ingredient, as each of them will give the dish that extra touch.

Agnieszka Wykowska What are the most important (for you) lessons learned?

Davide De Tommaso Certainly the most important are about teamwork.

Personally, I find nothing more satisfying and effective than collaborating together towards a shared goal. But you know, this is not easy, and usually requires considerable personal effort. I believe that the key is to be aware that teamwork is an opportunity for professional development and, above all, for personal growth.

It is a source of pride to be part of cutting-edge research, even if it can often be frustrating for several reasons. I believe that being part of a team where you feel recognized for your identity, for your abilities, gives you the strength to carry on in difficult times.

I learned this the hard way, and I am convinced that this is the best soil to sow your ideas, and find your path more quickly.

Agnieszka Wykowska What would be the message to the robotics community that you could bring based on this work?

Davide De Tommaso Typically, a technician approaches a robot more or less as s/he would approach a washing machine, or anything else equipped with sensors and actuators.

However, this does not apply to the majority of people who have only seen a robot in movies at the cinema, or read about it in a science fiction novel. The moment it shows, or behaves, in a human-like way, they'll anthropomorphize it. They won't ask you for an instruction manual to make it work, but instead they will become attached, they will empathise, they will be disturbed and affected by it.

As a roboticist, never make the mistake of underestimating the expectations and ability to attribute intentionality to a human-like robot robot by users.

Agnieszka Wykowska Would you ever consider iCub as an intentional agent?

Davide De Tommaso After all these years of working in close contact with cognitive scientists, I've probably become quite influenced, but I think there's something special about iCub. You know, I programmed its human-like behaviours many times, I even used distorted recordings of my own voice to make it talk. But for some strange reason, every time I see iCub laugh, it makes me laugh too.

8.2 Summary of the Take-Home Messages

Our empirical programme that addressed the theme of the Intentional Stance towards humanoid robots has resulted in numerous interesting take-home messages on when people might potentially treat robots as intentional agents. First, we have learned that 'intentionality' is much in the eye of the beholder. Indeed, our eyes have internal 'lens' (literally and metaphorically) that make us perceive the world in an individually biased manner. This lens/bias is coloured (metaphorically) with different crayons for each individual person—one crayon is education, other crayons are personality and individual predispositions, knowledge and information that is at the individual's disposal, cultural background (and many more). Through such coloured lens we perceive the world and others around us. Therefore, whether we perceive a robot as an intentional agent, whose actions are a result of its own intentions, desires, and goals; whether we see it as our potential social companion; or whether we see it only as an artefact (stuff made of metal and wires) is, to a large degree, 'on us', and on the biases with which we approach that specific robot. Our work presented here, together with the work of other researchers, who have addressed this topic, showed that length of education, the type of education, as well as background knowledge (when consolidated), influence the likelihood of adopting the Intentional Stance. Cultural differences might also play a role, especially in the broader context of anthropomorphism. Individuals embedded in anthropocentric cultures might relate to a human when being asked the question about the robot's mind, while non-anthropocentric cultures might rather think of agentic capacities, without referring to the human.

Most interestingly, when looking at individual biases at the neural level, it is possible to identify a pattern of neural activity at rest that might determine what attitude/stance one might take towards a robot. Thus, a given state of mind, at a given moment, might determine how we engage in an interaction with a robot. Something that is true probably for any social interaction—but could be used by roboticists to tune the robot and adapt its behaviour to a momentary mental state of the user, thereby making the robot better attuned to the human interaction partner.

Having said this, it is obviously not only the human observer's biases that play a role in the degree to which the Intentional Stance is adopted. The robot's appearance and behaviour are key factors that might elicit a certain impression on the human, which, in consequence, affects the likelihood of adopting the Intentional Stance. Human-like physical features (face, eyes, arms, and legs) not only activate the 'human' and 'like-me' representation, but also display a human-like (and thus, familiar) motor repertoire. If a robot has a motor repertoire similar to that of one's own, then the movements and actions can be presumably explained in a similar manner to those performed by ourselves. This might be evoked through the activation of the motor resonance mechanism (through the mirror neuron system), which has been shown to respond also to robot actions, as long as they resemble the human motor repertoire. This brings me to the next major factor crucial for adoption of the Intentional Stance, namely human-like behaviour (movement kinematics) and biological motion in general. Movement cues are extremely potent in interpreting the behaviour in intentional terms, and this has been known in literature since the classical work of Heider and Simmel (1944), who showed that even simple geometrical figures can elicit interpretations of their behaviour with reference to mental states. Designers of animated characters are very well aware of the importance of human-like behaviour for making the characters appear as intentional agents, to whom an observer attributes a whole plethora of thoughts, feelings, and emotions. And so are designers of robots—also in robotics, researchers, designers, and entrepreneurs pay attention to making robots appear human-like in their behaviour. As Prof. Ishiguro mentioned in the interview included in this book: 'While giving a robot a human-like appearance is beneficial for sensing intentionality, sometimes appearance is not as crucial as interactive behaviours'. Unfortunately, as discussed in Chapter 4, it is a much more difficult task when working with a physically embodied robot than it is with animated cartoons. This is all due to the fact that the physically embodied robots have very different hardware constraints than the human body.

Finally, the interaction per se ... Our research has shown that various aspects of interaction with the robot are as important as the robot characteristics themselves. After an engaging interaction with the robot, when participants might have the impression of having shared experiences with the robot, being a 'team', or, even better, of having shared goals and tasks, participants might be much more likely to increase the likelihood of adopting the Intentional Stance towards that robot. This might be presumably due to having the impression of the robot being 'like me', as they shared the same experience and responded to it similarly. Being 'like me' activates the representation of a 'human' in the observer's mind and, since a human is an intentional agent, so should be the robot. Important to note here, however, is that the research on 'Interaction' reported in Chapter 5 was all conducted in laboratory settings where the interactions were rather short and the 'illusion' of sharing experience could be maintained. Obviously, for longer-term interactions, to maintain the impression of sharing experience, or long-term goals, the robot could not 'pretend', as it did in our experiments. Perhaps for creating the impression of sharing experience in a long-term interaction in the real world, the robot would need to be genuinely endowed with mental states and intentionality, and thus become a 'true believer' (a genuinely intentional agent), in Dennett's terms (Dennett, 2003).

8.3 Conclusions: Intentional Stance as a Gradient

Having considered all the results that have emerged from the 6-year project on the Intentional Stance towards humanoid robots, an ultra-short summary of the take-home messages would be: ***People are inclined to adopt the Intentional Stance towards robots to some extent (even if robots do not genuinely possess mental states). This extent depends on the observers themselves, on their individual biases, on the robot's appearance and behaviour, and on the interaction between the two***. However, let's not forget the word 'extent' in this context.

Indeed, the Intentional Stance seems to be a gradient, rather than a binary (on/off) mechanism. For example, one can adopt the Intentional Stance to the lowest possible degree by using only 'intentional vocabulary' as a manner of simplified communication (but not really explaining or predicting behaviour of a system in intentional terms). We tend to use such manner of speaking when we say 'my computer does not want to restart'. Obviously, we are perfectly aware that the computer does not have the will (or lack thereof) to restart.

The higher degree of adopting the Intentional Stance would be in cases of suspension of disbelief—when we watch cartoons—we do explain and predict the behaviour of Donald Duck with reference to mental states, but that is because we 'pretend play' together with the authors of the cartoon, for the sake of enjoying the story and the narrative.

A yet higher degree of adoption of the Intentional Stance is when we explain and predict the behaviour of a system with reference to mental states, but we are agnostic regarding the capacity of the system to possess mental states. This might be the case when we do not understand how the system works, and Intentional Stance might be the most accessible strategy we can use. This indeed might happen in relation to robots and AI in general.

Finally, the highest degree of adopting the Intentional Stance is when we not only predict and explain behaviour of a system with reference to mental states, but we also truly attribute mental states to the system.

8.4 Open Questions for Future Research

Obviously, an entire research programme dedicated to a topic, no matter how long-lasting that programme is, generates more open questions than answers. Some of our research left us with the open question regarding the dissociation between implicit and explicit measures. In Chap. 4, we described some results (Ghiglino et al., 2020) where implicit measures of 'attunement' with the robot showed a different pattern than explicit reports. As is often the case in psychology, perhaps our brain implicitly responds (or attunes) to a certain robot/behaviour/interaction, even though this response is then 're-interpreted' differently at a higher, more explicit level. Thus, addressing the topic of the Intentional Stance with the use of both implicit and explicit measures is crucial and might provide a more comprehensive picture of the topic. Most importantly, future research should address the question of where this dissociation comes from, why does our brain 'reinterpret' its own initial responses, and how does this relate to the stance we take towards robots at the explicit, conscious level.

A further open question for future research is the relationship between social attunement and the Intentional Stance. Does the Intentional Stance evoke social attunement, or is the direction of the relationship opposite, meaning that social attunement evokes the adoption of the Intentional Stance? So far, our results showed bidirectionality of this relationship, as, on the one hand, Marchesi et al. (2022) showed that socially attuning with a robot through, for example, shared experience, increases the likelihood of adopting the Intentional Stance. On the other hand, Navare et al. (2024) showed that increasing the adoption of the Intentional Stance increases social attunement (in the form of sense of joint agency in a joint complementary task). This bidirectionality of the relationship between the Intentional Stance and social attunement is quite plausible, as social attunement might evoke the impression of the robot being 'like me' and thus being an intentional agent (social attunement ➔ Intentional Stance); while adopting the Intentional Stance might also lead to higher social attunement, as we might treat intentional agents more as in-group members, and thus, we might be more inclined to socially attune with them (Intentional Stance ➔ social attunement). Although plausible, and having gained some initial empirical support, the bidirectionality of this relationship should be further systematically tested.

When discussing the implications of adopting the Intentional Stance, one cannot omit the debate about ethical and moral implications, as discussed in the Introduction and in Chapter 7. It is of great importance to understand whether a moral stance towards a robot presupposes the Intentional Stance, and whether robots should be treated as *persons* in our legal systems, even if they are not true intentional systems. If the Intentional Stance is a necessary condition for a moral stance to be assumed, and if we would like humans to assume a moral stance towards robots (would we?), then we should design robots that do evoke the Intentional Stance. This is a highly debatable topic and should be the focus of future research in the context of social robots entering our societies. What are the risks and benefits of assigning moral responsibility to robots? What are the risks and benefits of designing robots that evoke the Intentional Stance? On the one hand, this might be beneficial for social attunement, which might be also beneficial, for example, in elderly care. On the other hand, we do not want people to be misled into attributing to robots' qualities that the robots do not have. For now, these themes are certainly open questions and should be addressed with great care in future research, in academic as well as political debate.

With this in mind, let me invite the reader to the final part of the book, the **interview with Prof. Daniel Dennett**, who was the originator of the concept of the Intentional Stance. On the 27th of January 2024, I had a unique opportunity to discuss with Prof. Dennett the topics of societal implications of creating robots that evoke the Intentional Stance.

Interview with Prof. Daniel C. Dennett

by Agnieszka Wykowska

Certainly it is possible, in principle, to make artificial agents, to make robots, that are just as fully intentional systems as we are, that have beliefs just as robustly as we do, and that can express them and debate them and defend them.

– Prof. Daniel C. Dennett

Photo credit: Alonso Nichols

Prof. Daniel Clement Dennett was a philosopher, writer, and cognitive scientist, renowned for his contributions to the philosophy of mind, philosophy of science, and philosophy of biology. Prof. Dennett was the Director of the Center for Cognitive Studies and Austin B. Fletcher Professor of Philosophy at Tufts University.

Among many widely read and acclaimed works in philosophy of mind and philosophy of science, Prof. Dennett has authored the seminal books *Brainstorms: Philosophical Essays on Mind and Psychology* (MIT Press, 1981), *Consciousness Explained* (Little, Brown and Co., 1991), *Darwin's Dangerous Idea* (Simon & Schuster, 1995), *Freedom Evolves* (Viking Penguin, 2003), and, most importantly for this conversation, the *Intentional Stance* (MIT Press, 1987). Prof. Dennett is recognized as a prominent figure in atheism and secularism among the 'Four Horsemen of New Atheism' alongside Richard Dawkins, Sam Harris, Christopher Hitchens. Throughout his career, Prof. Dennett's innovative insights have significantly enriched our understanding of the mind and its complexities, cementing his legacy as a leading thinker in contemporary philosophy.

Summary

On the 27th of January 2024, Prof. Agnieszka Wykowska met with Prof. Daniel Dennett for a conversation (online) on the topic of the Intentional Stance towards humanoid robots. Prof. Wykowska summarized to Prof. Dennett the main theme of her ERC project 'Intentional Stance for Social Attunement' and the contents of this book. Then, they discussed the questions of whether humanoid robots can indeed be considered true intentional agents, if the Intentional Stance is a gradient, and the moral consequences of designing robots that might evoke the Intentional Stance in vulnerable individuals. This conversation concludes this book that is centred around the concept of the Intentional Stance—a concept that originated in the works of Prof. Dennett.

Keywords
Intentional Stance, Intentional agents, True Believers, Design Stance

Conversation on the Intentional Stance Towards Humanoid Robots

Agnieszka Wykowska Thank you very much for accepting my invitation and for your time today. I just wanted to ask you a few words about your thoughts on the Intentional Stance. I would start by asking: do you think that humanoid robots could ever become True Believers? And do you think the Intentional Stance will ever be an efficient strategy to adopt towards artificial agents?

Daniel Dennett Oh, yes, absolutely! Because they won't differ in this regard from us; in some ways, they will be like animals, like dogs and cats, or predators like lions and tigers, or animals that are trying to escape us. And we will need to use the Intentional Stance to trap them or recover control over them, if they do escape from us. Those applications of the Intentional Stance, which come naturally to us, will be just as well grounded, as they are for us.

Agnieszka Wykowska Do you think that those artifacts will actually develop mental states? They would actually be intentional agents?

Daniel Dennett Yes, I think that's what being an intentional agent is. Let's go way back to one of my inspiring moments when a designer of a chess-playing program said about a rival chess program that "it thinks it should get its queen out early". That was an Intentional Stance attribution to that program. Everybody in the room knew that he did not mean that it had a sentence in its head that meant "I should get my queen out early" but they also knew that this was a deeply diagnostic and accurate betrayal of a feature of this system.

That's the Intentional Stance in action.

That was a crude example and it was an early chess playing computer, but it goes all the way up. It goes for insects and people, there is no sharp dividing line. Although with human beings, the fact that we can articulate with language and can express our beliefs and our opinions (note, I make a distinction between beliefs and opinions) makes a difference. And the fact that human beings, because of their linguistic capacity, ***articulate*** beliefs and desires, this is an important difference. I wish I had developed this further in my work.

"How do you get from beliefs to opinions?" You get from beliefs to opinions by saying what you are willing to say. You're betting on the truth of sentences.

Agnieszka Wykowska So what you are saying is in line with the idea that the Intentional Stance is a gradient, rather than a binary process. I certainly share this view, after having reasoned about the Intentional Stance for a while, and based on the results we have obtained throughout our project.

Daniel Dennett It's a gradient. Computer scientists themselves often use intentional language. Let's say… "you want to make your computer think it's printing to a printer when in fact it's just printing to a USB stick". Of course that's very metaphorical, but also a very useful metaphor: if everyone is on the same page with using the language, then it works out fine. And in fact, the programmers and the like, they communicate very effectively with each other, using this intentional language. So it's a very useful part of their own sophisticated user illusion of the computers that they are programming.

Agnieszka Wykowska If it's a gradient, do you think there will still remain some sort of a difference between adopting the Intentional Stance towards humans versus the Intentional Stance towards sophisticated robots?

Daniel Dennett It all depends on the artificial agents.

That is why many of the arguments about large language models (LLMs) are misguided because they are asking whether that mystical extra feature (consciousness) is really there. There is no mystical extra feature, it is just "better and better" and "more and more". And when you've got a large language model that you trust, if you do, you probably shouldn't [laugh], then that's just it!

The hunt for the dividing line is a misguided hunt. We should recognize that there is no such dividing line; but we want to maintain a line in our societies, socially and legally, because we want to keep these things under our control. We don't want them to become "artificial colleagues", we want them to be just work machines. But that line itself is only a gradient.

Agnieszka Wykowska Absolutely! I remember when we met in Bochum in 2022, you had some reservations toward building systems in such a way that they would elicit the Intentional Stance in the user. Where do these reservations come from?

Daniel Dennett Well, most human beings, those who are not hi-tech or internet gurus or computers scientists or programmers, follow the Intentional Stance rather naively. And they put more trust in the systems than the systems deserve. That is because the rationality assumption is built right into the Intentional Stance, so people overestimate.

A principle of charity is built right into the naïve, and automatic, irresistible adoption of the Intentional Stance. It carries with it a level of trust, which is often too much.

Agnieszka Wykowska Trust in what sense?

Daniel Dennett Trust that you are being understood, and trust that the protestations of the thing you are talking to are sincere and are to be trusted. That is: Are they speech acts? Well, they're like speech acts, coming from the LLMs. They're sort of fake speech acts. They're like speech acts of a man in the imitation game pretending to be a woman and trying to win the imitation game. You want to be careful about trusting somebody who can do that so well.

Agnieszka Wykowska What do you think are the dangers of over-trust? Let's think of a social robot taking care of an elderly person, and the interaction is smooth and very social. The person believes the robot understands them. Why is that dangerous, if it serves its purpose?

Daniel Dennett Well, we want to know. Let me put it this way. Suppose, instead of a robot you put a human caregiver in that place. One thing you want to know is if the human caregiver is not being indirectly controlled by somebody else.

Let's make up a little story. John has an uncle who is very rich and John wants the uncle to give his fortune to him (to John) when he (the uncle) dies. Therefore, he wants the uncle to change his will. So John goes to a wonderful nurse maid who is taking care of his uncle Bill and pays her big money to shift the uncle's allegiances to John. She is very good at it, and pretty soon the uncle calls in his lawyer and changes the will leaving all the money to John.

Bad? Yes, bad!

Well, we replace the human caregiver with a robot and the same situation exists. Now the question is who does the AI caregiver care the most about? Itself? Its corporation that designed and built it in the first place? What if Uncle Bill starts buying stocks in that company? [laugh]

Agnieszka Wykowska So it's a matter of responsibility - who is the one eventually responsible for the morally bad outcome.

Daniel Dennett Yes, yes.

> ***So if the caregiver is not human, then the caregiver is in a certain sense disabled for moral responsibility. Because it does not have the mortality and fragility that you and I do. It can be backed up and brought back to life. It does not share with us the fragility that we normally assume. That is very important.***

Agnieszka Wykowska Right. But again, fragility is also a gradient. Also humans have different degrees of fragility. Less and more fragile and less and more manipulative and less and more responsible.

Daniel Dennett Yes, exactly. In fact, lets highlight this with a little example. I am just making this up as we go along. Imagine you have an elderly relative (a grandmother) and you have a choice between two different caregivers. One is, apparently, a gentle poor person who does really need the money and who is willing to make tremendous sacrifices for your grandmother. The other one is a son of a billionaire, and he is doing this for fun. Which one do you want to take care of your grandmother?

Agnieszka Wykowska So basically, if I am to extend this to the robot case, what you are saying is that robots just simply inherently do not care.

Daniel Dennett Not if we don't make them so that they care.

> ***Can we make them [robots], so that they care? Can we make them so that they can worry about their own death? We could. And that would be very dangerous.***

We don't want to make them that autonomous, we want to make them so they are—as Joanna Bryson says—just slaves. So that we have no moral obligation to them, but they do obey. Can we make them like that? Yes, we can. The thing is not to make them more autonomous than that.

Agnieszka Wykowska Very interesting! I don't want to keep you much longer. But I wanted to ask you a final question. If you were to give some advice to future researchers in robotics and Human-Robot Interaction, what would you advise in terms of robot design. Where do we want to go in robotics an AI?

Daniel Dennett Well… let's put it this way…

> ***We want to have something like a watermark system. That means that a grandma always knows that she is dealing with a robot and not with a human being, not with a person.***

I know there will be a very strong economic temptation. Eldercare is going to be a big area, I think, because eldercare is not a very good life for human beings [i.e., nurse maids]. Or like telephone operators: Nobody laments the loss of hundreds of thousands of jobs of being telephone operators. And being an eldercare nurse maid is not a very good life. So we would like to replace that with robots. There are better things for people to spend their life doing. But then we also want to make sure that the human touches that we build in (which is not that hard) are recognized as such [as artificially built-in] and limited. This is a tricky point.

In the movie "HER", the protagonist falls in love with his companion. That is a realistic scenario for the future. We are going to have grandmothers and grandfathers who fall in love with their robot caregivers. Is this ok? Only if this is not an exploitation of the person.

Agnieszka Wykowska Thank you very much for your most insightful comments, and for all the priceless inspiration you have given to my team to pursue a project on this exciting topic. It's been an incredible honor to talk to you today.

Daniel Dennett Thank you, it's been fun talking with you.

Prof. Daniel Dennett and Prof. Agnieszka Wykowska. (Photo taken by Dr. Serena Marchesi on the occasion of a workshop 'To be or not to be … conscious. Phenomenal Realism and Illusionism', organized by Dr. François Kammerer and Prof. Tobias Schlicht, Bochum, 2022)

References

Dennett, D. C. (2003). True believers: the intentional strategy and why it works. In T. O'Connor & D. Robb (Eds.), *Philosophy of mind: Contemporary readings* (pp. 370–390). Routledge.

Ghiglino, D., De Willemse, C., Tommaso, D., Bossi, F., & Wykowska, A. (2020). At first sight: robots' subtle eye movement parameters affect human attentional engagement, spontaneous attunement and perceived human-likeness. *Paladyn, Journal of Behavioral Robotics, 11*, 31–39.

Heider, F., & Simmel, M. (1944). An experimental study of apparent behavior. *American Journal of Psychology, 57*, 243–259.

Marchesi, S., De Tommaso, D., Perez-Osorio, J., & Wykowska, A. (2022). Belief in sharing the same phenomenological experience increases the likelihood of adopting the Intentional Stance towards a humanoid robot. *Technology, Mind and Behavior, 3*(3), 10.1037/tmb0000072.

Navare, U. P., Ciardo, F., Kompatsiari, K., De Tommaso, D., & Wykowska, A. (2024). When performing actions with robots, attribution of intentionality affects the sense of joint agency. *Science Robotics, 9*, eadj3665.

Index

A. Wykowska (ed.), *Intentional Stance Towards Humanoid Robots*, https://doi.org/10.1007/978-3-031-65483-1

MIX
Papier aus verantwortungsvollen Quellen
Paper from responsible sources
FSC® C105338

If you have any concerns about our products,
you can contact us on
ProductSafety@springernature.com

In case Publisher is established outside the EU,
the EU authorized representative is:
Springer Nature Customer Service Center GmbH
Europaplatz 3, 69115 Heidelberg, Germany

Printed by Libri Plureos GmbH
in Hamburg, Germany